Sights on Jarvis

No. 1 Bombing and Gunnery School, 1940-1945

by Robert Schweyer

To Gerry
with best wishes!

Robert Shwy

Nov 1, 2003

Heronwood Enterprises

Copyright © **Robert D. Schweyer**, 2003

Editor: Cheryl MacDonald
Design and production: Heronwood Enterprises
Printing: Morris Printing Services Inc., Simcoe, Ontario

Front cover: Student air gunner with a Vickers machine gun, No. 1 Bombing and Gunnery School, July 1941. NAC WRF-402/Nicholas Morant

Published by
Heronwood Enterprises
R.R. 2, Nanticoke, Ontario, N0A 1L0
www.heronwoodent.ca

National Library of Canada Cataloguing in Publication
Schweyer, Robert D. (Robert Delmar), 1953-
Sights on Jarvis : No. 1 Bombing and Gunnery School, 1940-1945 / Robert D. Schweyer.

ISBN 0-9694896-3-3

1. Canada. Royal Canadian Air Force. Bombing and Gunnery School, no. 1. 2. Bombing, Aerial—Study and teaching—Ontario—Jarvis. 3. Aerial gunnery—Study and teaching—Ontario—Jarvis. 4. British Commonwealth Air Training Plan.

UG635.C22J37 2003   358.4'0071'071336   C2003-905574-4

Contents

OTHER THINGS

Oh! he has hastened after other things,
   And traveled far.
The pendulum slow, slower swings
   On his last ride,
   And in exchange
For visions glimpsed of further earthly years
   He sought a guide,
   Nor thought it strange
   To choose a dream.
A dream most fair, and brave
With the sweet peace that brings
The quiet answer to all other things.
Beyond a war-scarred sky
   He signaled his reply,
     And gave
His battered aircraft to the guiding beam
That takes the pilot to another shore
   Where many mansions are.
Bright lights of home, an unlatched door
 And happy landings evermore.
   On angel wings,
   And, other things.

Written by Rose Whitehead of Kaeo, New Zealand upon receiving word of
the death of her son John William in a flying accident at Jarvis, August 18, 1942.
Reprinted by kind permission of the family.

Introduction

Looking back, I suppose I really started this book at the age of eight, one spring day in 1961. That was the year my boyhood chum Wade King and his dad Howard gave me my first real piece of aircraft wreckage—a small, tattered bit of fabric with the date and name of an RCAF guard pencilled on the back. On that day, they led me to a field at the edge of their bush, and more precisely, to a large circular depression quite visible in the ground. Howard went on to explain that, 20 years earlier, he and his father were working ground with a team of horses when a silver-coloured plane came screaming down and slammed into that very spot. When asked what happened to the pilot, Wade's dad just looked away and said he had been killed. We could tell it still bothered him.

I remember standing there for the longest time, just staring at the spot where only a few weeds sprouted unevenly from an otherwise healthy field of clover. I imagined what it must have been like—the fatal dive, the tremendous crash that followed and Howard and his father trying to calm the terrified horses mere yards from where we now stood. After being shown some additional bits and pieces of the plane unearthed over the ensuing years and now stored in a six-quart basket in the King garage, I jumped on my bike and raced home to tell my parents what I had learned and to show them my latest acquisition.

My dad knew all about it. He too had been working in a field on our farm that afternoon and had heard, then seen, the plane go down. He told me it was a Yale from Dunnville. He then went on to tell me about the two Ansons which got caught in a freak snow squall just before church one Sunday morning the following year. He pointed to our bush and explained how he watched the planes flying one behind the other just above the treetops, headed west and barely visible through the driving snow. One of the two pilots made it safely back to the airport he told me, the other had not.

What was a Yale? I wanted to know, and what were Ansons? Who was flying them and what were they doing here? I learned much later that the Anson which crashed in the storm was piloted by a 19-year-old American from New York City and the Yale at the King farm by 21-year-old Frank Schwartz of Montreal. Both were pilots in training and both were flying solo at the time. I now have their pictures, and of course, I still have the piece of fabric.

It was hearing stories like this over and over while growing up in that part of Ontario that intrigued me. I became absolutely determined to learn as much as I could about what had happened in my own backyard years before I was born. As a youngster, I read everything I could find about the Royal Canadian Air Force and quickly became an expert in aircraft recognition at a time when most kids my age were fishing and playing baseball. I talked to neighbours and relatives at every opportunity. Everybody, it seemed, had a story. It was while attending Niagara College in Welland in 1971 that I had the good fortune of having Hugh A. Halliday as one of my instructors. Hugh, now a well-known aviation author and historian, was formerly with the Historical Branch of the Department of National Defence and was able to steer me in the right direction for unit diaries and other official records in Ottawa. I made several trips to the nation's capital in the next few years and never looked back.

I continued to meet with anyone and everyone, taking notes and spending hours poring over old files at newspaper offices in Jarvis, Port Dover, Simcoe and Hamilton. That was 30 years ago. I took an active interest in the British Commonwealth Air Training Plan (BCATP) and more specifically, the wartime RCAF stations which surrounded my parents' farm. Per capita, this part of Canada had possibly the highest concentration of aerodromes in the country, with three major air schools located within a 30 mile-radius at Jarvis, Dunnville and Hagersville, and two additional relief fields at Kohler and Dufferin. I became fascinated with all of these, and in particular, the former airfield at Jarvis known as No. 1 Bombing and Gunnery School. As the first station of its kind in Canada, Jarvis initiated and perfected the various bombing and gunnery techniques ultimately used by the other ten such schools established under the BCATP. From August 1940 to February 1945, the station was regarded as one of the finest in Canada, cranking out more than 6,500 bomb aimers, air gunners and wireless operator/air gunners.

The British Commonwealth Air Training Plan was an agreement signed on December 17, 1939 between the governments of Canada, Great Britain, Australia and New Zealand whereby thousands of airmen from the British Empire would receive their training in Canada. Almost overnight airfields sprung up from Charlottetown, Prince Edward Island to Vancouver, British Columbia and by the time the plan was terminated in March 1945, a total of 131,553 aircrew had received their wings. Much has been written about the BCATP in recent years and its story is well documented elsewhere. *Sights On Jarvis*, therefore, is not intended to serve as yet another history of this scheme but rather, of one of the many schools that came into existence because of it.

At any airfield in wartime, accidents were inevitable and throughout the text you will read of the tragic events surrounding the deaths of most of the 40 young men killed at Jarvis. Without exception these accounts are quoted directly from official RCAF court of inquiry transcripts, which only recently were released for public information by the National Archives of Canada. Out of respect for the surviving wives, brothers and sisters, I have attempted not to include the usual hearsay evidence gathered at the scene or the graphic details often contained in medical officers' reports. My family and I feel privileged to have made so many friends among the relatives of those killed, most of whom have never had the opportunity to visit their loved ones' graves here in Jarvis. Tracing family members has been painstaking and has led to untold numbers of letters and telephone calls to all parts of this country, Australia, New Zealand, England and the United States.

Since the beginning, it has been my intention to provide readers with lots of pictures that best showed what life was like at a bombing and gunnery school. Included in the text, therefore, are numerous photographs obtained mostly from veterans' albums and never previously published. If one picture is worth a thousand words, they can tell you more than I could ever hope to.

In conclusion, I have thoroughly enjoyed researching this book for the past 42 years! If I have but one regret, it is that I did not finish it sooner. Many of those who influenced me so much, including my father, are gone now and I only hope this book does justice to the ones who remain.

Robert Schweyer
Jarvis, 2003

Acknowledgements

It would be impossible to list here absolutely everyone who has shared their personal experiences or pictures with me over the years. I would however like to take this opportunity to acknowledge at least some of those individuals and organizations who contributed information and photographs for this book, beginning with the following veterans of No. 1 Bombing and Gunnery School:

Ted Potter, Beamsville, Ontario; Edmon Ryerse, Port Dover, Ontario; Harvey Ohland, Simcoe, Ontario; Norm Williams, Hamilton, Ontario; Ben Bohan, Selkirk, Ontario; Jack Paton, London, Ontario; Tom Lawrence, Fort Erie, Ontario; R. Duke Waddell, Rancho Mirage, California; Howard Hewer, Toronto, Ontario; Irene (Goldspink) Miller, Dunnville, Ontario; George Sibley, Brechin, Ontario; Bruce Garber, Fisherville, Ontario; Ethel (Davis) McNeilly, Hagersville, Ontario; Ian Moon, Hamilton, Ontario; Roy Thorne, Simcoe, Ontario; James Falconer, Chilliwack, British Columbia; Harry Dean, Bowmanville, Ontario; John C. Milner, Port Dover, Ontario; Dick and Doris Hennessy, Cape Neddick, Maine; Norm Lillico, Ottawa, Ontario; Albert Whyley, Woodstock, Ontario; William P. Armes, Toronto, Ontario; Amy (Stanford) Evans, Mississauga, Ontario; A.L. d'Eon, Don Mills, Ontario; Les McLean, Cabri, Saskatchewan; Art Rimmer, Brampton, Ontario; John St. Thomas, St. Louis DeKent, New Brunswick; Tom Kernaghan, Hamilton, Ontario; Jessie (McGregor) Brown, Norwood, Ontario; Herbert Hacker, Leamington, Ontario and Jack Doerr, Exeter, Ontario.

A significant number of photographs and documents were also made available to me by the National Archives of Canada; Canadian Forces Photographic Unit, Department of National Defence, Ottawa; Directorate of History, Ottawa; John

Lutman, London Free Press Collection of Photographic Negatives, University of Western Ontario Archives, London; Canadian Aviation Historical Society, Toronto Chapter, Willowdale, Ontario; Canadian Harvard Aircraft Association, Tillsonburg, Ontario; No. 6 SFTS Reunion Association, Dunnville, Ontario; Jarvis Public Library; Canadian Warplane Heritage Museum, Hamilton, Ontario; Department of Defence, RAAF, Canberra, Australia; Bill Fess Collection, Jarvis, Ontario; Elgin County Military Museum, St. Thomas, Ontario; Mick Henning, Port Dover, Ontario; Frank and Betty Scholfield, Dunnville, Ontario; Enid (Williams) Blume, Simcoe, Ontario; Wing Commander A. Judd Kennedy (Ret'd), Toronto, Ontario; Joan Stephens, Brookfield, Nova Scotia; Jeanne Russel, Simcoe, Ontario and Robert Scholefield, Westmount, Quebec.

Several other individuals have contributed substantially in other ways to this book. I am indebted to Rosemary "Greg" Hilton and Roberta Chapman of the Jarvis Public Library for making all their resources available to me; Wally Fydenchuk of Crediton, Ontario, No. 9 SFTS historian, who has continually supplied me with names, addresses and photographs over the years; and to Adam Wlodarczyk, Burlington, Bob Finlayson, Hamilton, Rick Radell, Toronto and the Kindurys family at Studio One Image Centre, Simcoe, for photo reproduction. A special thanks to the members of the London South Rotary Club, London, Ontario for their generous donation of a computer and to Adam Kowalsky of Simcoe, currently attending the University of Western Ontario, for his ongoing efforts to make me (somewhat) computer literate.

During the course of researching this book, I have had the pleasure of corresponding with several relatives of those killed at Jarvis. Family members sharing their personal memories, pictures, letters and log books with me include Lynn and Cath Whitehead, Auckland, New Zealand; Nell Day, Whangarei, New Zealand; Neil and Deci McNabb, Griffith, N.S.W., Australia; Dr. Burdge Green, Houston, Texas; Patricia Troutbeck, Seaforth, Ontario; Lillian McCrank, Kirkland Lake, Ontario;

Bill and Isobel Johnson, Simcoe, Ontario and Lois Foulds, Martintown, Ontario. To all, thank you for permitting me to intrude into a part of your past you would sooner forget.

Thanks also to Ontario residents Stan Morris, Port Dover; Charlie Cox, Simcoe; Myrtle Johnson, Jarvis; Dave Otterman, Port Dover; Jim Miller, Jarvis; Don Scruton, Ancaster; Larry Hare, Stoney Creek; Kirk Brown, Hagersville; Steve Mouncey, Peterborough; Vic Hare, Selkirk; Chuck Bartram, Burlington; Roy Barnard, Selkirk; Chris Smelser, Fisherville; Clare Dennis, Port Dover; Robert Hare, Selkirk; Earl Fleming, Delhi; Glenn Link, Kohler; Ken and Georgia Painter, Port Dover; Tom Millar, Turkey Point; Shawn Barnard, Selkirk; Vic and Janet Lytle, Cambridge; Thelma Swent, Rainham Centre; Ruth Pond, Jarvis; Paul Morley, Hamilton; Hartley Garshowitz, Thornhill; Bob Hall, Cayuga; Norm Shrive, Burlington; Jim Burnke, Ailsa Craig; Gord Hill, Dundas; Peter Robertson, Ottawa; Jack Cooke, Jarvis; Hugh A. Halliday, Orleans; Howard Elliot, Jarvis; Orval Fletcher, Selkirk; Bruce and Elizabeth Milner, Port Dover; Lawrence Jonathan, Ohsweken; George Jepson, Jarvis; Loys Readwin, Guelph; Carl Colley, Ancaster; Trevor Meldrum, Greensville; Bill Cumming, Jarvis; Jack Evans, Hamilton; Floyd Smelser, Dunnville; Geraldine Melenbacher, Hagersville; Carl Vincent, Ottawa; John Reid, Hagersville; Bob Fugard, Burlington; Jack Dennis, Port Dover; Mary Miller, Jarvis; Harold Waterbury, Fisherville; Frank Nelson, Dundas; Florence Mussel, Port Rowan; George Smith, Port Ryerse; Ruth (Ralston) Stuart, Hamilton; Erik Roggenkamp, St. Catharines; Margaret Roberts, Clanbrassil; Don Mattice, Selkirk; Jack Culver, Rainham Centre; Carman McMullen, Ancaster; Doug Hare, Simcoe; Ron Passmore, Toronto; Wayne Ready, Hamilton; Jack Dundas, Ridgeville and Greg Burnard, London. Regrettably, this by no means includes everybody who has contributed in one way or another over the years and to those I have inevitably missed, my sincere apologies.

This book has been a family effort. I would therefore like to thank Cindy, Sarah and Matthew for their patience and understanding these past two years. Writing demands time and I have taken my lion's share of it in recent months. To their credit, all three have developed a keen interest in the people and events in this book and each has been of enormous help to me, often without even knowing it.

Finally, I wish to acknowledge my mother Blanche and my late father Delmar. It was they who fostered my interest in airplanes as early as 1955 when at the age of two, I would reportedly point skyward and exclaim, "Pwane... a pwane!"

Glossary

This list is provided for the convenience of those unfamiliar with aviation, military or World War II terms.

Ranks in the RCAF

Aircraftman, 2nd Class	AC2
Aircraftman, 1st Class	AC1
Leading Aircraftman	LAC
Corporal	Cpl.
Sergeant	Sgt.
Flight Sergeant	F/S
Warrant Officer, 2nd Class	WO II
Warrant Officer, 1st Class	WO I
Pilot Officer	P/O
Flying Officer	F/O
Flight Lieutenant	F/L
Squadron Leader	S/L
Wing Commander	W/C
Group Captain	G/C
Air Commodore	A/C
Air Vice Marshal	A/V/M
Air Marshal	A/M
Air Chief Marshal	A/C/M

Medals and Awards

Air Force Cross - awarded to officers and warrant officers for courage or gallantry while flying, but not in presence of enemy.

Bar - awarded to medal recipients for additional acts of bravery.

British Empire Medal - Order of the British Empire medal for commissioned, warrant and subordinate military officers.

Distinguished Service Order - Commonwealth medal for meritorious or distinguished military service in war.

Distinguished Flying Medal - Commonwealth medal for courage or devotion to duty while flying in the presence of the enemy. Awarded to non-commissioned officers and men.

Medals and Awards (continued)

Distinguished Flying Cross - Commonwealth medal for courage or devotion to duty while flying in the presence of the enemy. Awarded to officers and warrant officers.

George Medal - Commonwealth medal, usually awarded to civilians in wartime for acts of conspicuous bravery.

Victoria Cross - for conspicuous bravery. Highest military award in British Commonwealth.

Abbreviations

A/C	aircraft
BCATP	British Commonwealth Air Training Plan
CO	commanding officer
MT	motor transport
NCO	non-commissioned officer
RAAF	Royal Australian Air Force
RAF	Royal Air Force
RCAF	Royal Canadian Air Force
RNAF	Royal Norwegian Air Force
RNZAF	Royal New Zealand Air Force
SP	Service Police
USAAC	United States Army Air Corps
WAG	wireless air gunner
WD	Women's Division

Aviation and Military Terms

Air Ministry Bombing Teacher - (AMBT) British device which simulated bombing runs.

Central Flying School - hub of flying activity in Canada, located in Trenton, Ontario.

Armourer - ground crewman responsible for bombs, guns, ammunition and related equipment.

Astro navigation - navigating by stars.

Barrack block - sleeping quarters.

Bomb-bay - part of an aircraft where bombs were stored and released.

Central Navigation School - advanced navigation instruction school located in Rivers, Manitoba.

Crash tender - fire truck or high-speed boat equipped especially for responding to airplane accidents.

Dope - pasty liquid used to coat and protect the fabric covering of aircraft and to make repairs.

Drogue - cylindrical nylon target towed behind an aircraft.

Dual - accompaniment in aircraft by an instructor, or occasionally, another student.

Elementary Flying Training School - where student pilots flew their first aircraft.

Embarkation Depot - an airman's final posting prior to departing Canada for overseas duty.

Flight - name given to various courses or sections on a base to distinguish them; *e.g.*, bombing, gunnery, drogue, maintenance.

G-string - small wire attached to airman's harness to prevent his falling out.

Galt Aircraft School - where aero engine and airframe mechanics received training. Located in Galt (now part of Cambridge), Ontario.

GIS - Ground Instruction School - building(s) where students studied various theories and ground-related subjects in a classroom setting.

GO gun - Gas-operated Vickers machine gun.

Gosport tube - early communication system in aircraft.

H-hut - military building that from the air resembled an H.

Manning Depot - where all new recruits reported and were introduced to air force life.

Northwest Staging Route - chain of airfields linking Canada and the US to Alaska and the USSR. Used for ferrying aircraft.

Quadrant hut - small building located on the bombing range from where bomb hits were sighted and reported.

Relief field - secondary field where student pilots could practice take-offs and landings away from the main aerodrome.

Scrubbed - cancelled.

Service Flying Training School - where pilots were instructed in advanced flying.

Sortie - British term for mission. Also known as op or trip.

Target tug - aircraft which towed the drogue.

Technical Training School - chief RCAF technical training school for mechanics and other skilled tradesmen located at St. Thomas, Ontario.

Training Command - formed in Canada during World War II to oversee and direct all training operations at BCATP units. Divided into four commands, with headquarters in Toronto, Winnipeg, Montreal and Regina.

Very pistol - flare gun.

Victory Loan - campaigns promoting public purchase of war bonds to help finance the war effort.

Washed out - flying cancelled due to weather OR failure of student to pass course.

Western Air Command - command which controlled all RCAF units operating west of the Ontario-Manitoba border.

Wind tee - large, swinging metal object shaped like a T that indicated direction of wind from the air.

Wireless school - where student WAGs learned Morse code, signalling and use of radios.

Note on Capitalization

Grammar and usage, including capitalization, have changed considerably since the Second World War. Although aware of military and aviation traditions regarding capitalization (the style preferred by the author) the editor has opted for Canadian Press (CP) style. This is the style used by most Canadian newspapers and therefore most familiar to the majority of readers.

Chapter One
1940 - War Comes to Walpole

*F*or the residents of Walpole Township the fighting seemed a world away. Though six months had passed since Canada declared war, nothing much had changed in the quiet little farming community situated just north of Lake Erie, 35 miles south of Hamilton, Ontario. Then early one spring afternoon, a telegram arrived in the village of Jarvis informing Mr. and Mrs. Emerson Miller that their son James, 26, had been killed in action. Sergeant Miller had, in fact, become one of Canada's first casualties of the war when his Whitley bomber was shot down while returning from a reconnaissance flight over Germany on March 28. Miller had bailed out but his parachute failed to fully open.

Hundreds from surrounding Haldimand County and from as far away as Toronto turned out for the simple but solemn memorial service that followed at the Wesley United Church in Jarvis. Reverend Stewart East, pastor of the church, paid tribute to the local farm boy who had worshipped there and who had left for England a year earlier to join the Royal Air Force.

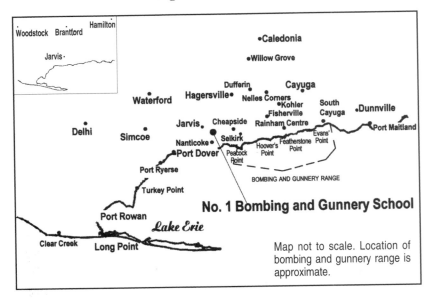

Map showing: Woodstock, Brantford, Hamilton, Jarvis (inset); Caledonia, Willow Grove, Dufferin, Cayuga, Waterford, Hagersville, Nelles Corners, Kohler, South Cayuga, Dunnville, Fisherville, Delhi, Simcoe, Jarvis, Cheapside, Rainham Centre, Port Maitland, Nanticoke, Selkirk, Evans' Point, Featherstone Point, Port Dover, Hoover's Point, Peacock Point, Port Ryerse, BOMBING AND GUNNERY RANGE, Turkey Point, **No. 1 Bombing and Gunnery School**, Port Rowan, *Lake Erie*, Clear Creek, Long Point. Map not to scale. Location of bombing and gunnery range is approximate.

"James Miller was obedient unto death," the minister began.

A few short months ago he heard the call. In those days fewer were the thoughts of war, but in the hearts of some of us was the conviction that there were forces shaping themselves in the world that could mean only one thing—the kind of tragedy that costs the lives of men.[1]

That Canada was indeed at war became even more apparent a week later when the *Jarvis Record* reported that the Department of National Defence had named Jarvis as the location of one of two bombing and gunnery schools to be constructed immediately as part of the massive British Commonwealth Air Training Plan. Its proximity to the lake meant an unlimited practice range over water for both bombing and gunnery exercises and this, together with the virtually flat characteristics of the surrounding countryside, made the area ideal. Moreover, the site selected had previously been developed as an airfield. As early as 1934 American Airlines operated an emergency landing strip five miles southeast of Jarvis, near the village of Nanticoke. It was here, with its single north-south sod runway, beacon light and radio post, that the government began purchasing additional property. Negotiations followed and eventually six farms comprising just under 600 acres were acquired for the paltry sum of $54,100.

Contractors from Toronto, Brantford and Windsor descended on the site. Farm boundaries disappeared as trees and fences were removed, ground levelled and graded, and cement for runways and foundations poured. Lumber arrived daily by the truckload as construction of six flight hangars, drill hall and some 45 H-huts and other buildings got underway. Local men looking for work showed up by the hundreds. One of them, 17-year-old Jack Dennis of Simcoe, arrived at the superintendent's shack on his bicycle one spring morning and along with dozens of others applied for a job as labourer. The superintendent, upon learning of the young man's exceptional carpentry skills, immediately set him to work laying floors and fitting trim around the hundreds of windows and doors taking shape in the many barrack blocks. Dennis, who was too young to enlist in the spring of 1940, recalled years later the deep sense of urgency impressed upon everyone to complete the project as quickly as possible. The war was reaching a critical stage and airfields like this one were badly needed, they were told. He spent several weeks on the job working ten hours a day, six days a week and earned $180 a month—a sizeable wage for a teenager in post-Depression Canada.

As construction of the main aerodrome forged ahead, work also proceeded on the bombing and gunnery range, which ran east along the Lake Erie shoreline from Peacock Point south of the airfield to Evans' Point in South Cayuga. Buildings

General view of the camp looking south. The building area alone covered almost fifty acres.
MICK HENNING

View of the school looking southeast in 1940. Several buildings have not yet been completed. JAMES FALCONER

there included a 40-foot observation tower, motor transport garage, boathouse and two quadrant huts for each bombing target anchored offshore. In addition, a 200-yard machine gun range, complete with moving targets and protective cement shelter, was established on property leased to the government at Hoover's Point, near Selkirk. Other work necessary to equip the range that summer included the building of three triangular bombing targets, a boat dock and a series of wooden splash targets used in air-to-ground machine gun practice.

The school's lone land target, located between the Third and Fourth Lines of the Six Nations Reserve. JACK PATON

Occupying an area of the lake 18 miles long by six miles wide, the entire range extended well into the water and covered more than 100 square miles. Cost of completing the range, including some 20 miles of interconnecting telephone lines, was $19,000.

By mid-summer construction was progressing well

1940 - War Comes to Walpole

Control tower, photographed from the rear cockpit of a taxiing Fairey Battle. The control tower offered unrestricted visibility of the field from four levels. At right is a view of the field from the tower. The aircraft are Fairey Battles. JARVIS PUBLIC LIBRARY

despite continuous rain. On July 25 an advance party of RCAF personnel arrived to find the heavy Walpole clay turned into an endless sea of mud. Walkways were non-existent and luckless was the airman who dared venture off the main roads after a thunderstorm. Daily parades were held on the tarmac each morning, but the airmen's nicely polished boots became so muddy getting there that any parade really was a travesty by military standards. Such conditions prompted Squadron Leader R.A. Cameron, chief instructor of the school, to coin the phrase, "You'll stick to your job in Jarvis," a slogan which stuck and later became the station motto. F/S F.C. Pearce, a member of the advance party, later recalled the hardships that existed during the summer of 1940:

> Officers, N.C.O's and men ate at a common mess hall ... Stew and rice were the favorite meals, served so often that some wag made a dummy of the messing officer and publicly hanged it, as a subtle suggestion. If the M.T. truck to Hamilton broke down, though, there wasn't even stew to eat. However, the canteen helped. You got service by poking your head through the window ...[2]

He remembered that for weeks following each meal, every airman kept and washed his own cutlery. Even the most basic requirements of service life were absent: shaving and washing had to be done using cold water trucked to the station daily

A student air observer and armourer check practice bombs prior to flight. Terminal velocity of the 11 1/2 pound bombs was 955 feet per second. DND PL3573

from Dunnville, 25 miles away. Pearce also recalled the first pay parade on August 15 when the paymaster ran out of funds after 20 men stepped forward and 15 others had to wait another eight days to draw their well deserved wages!

In the days that followed, the neighbouring communities of Jarvis, Port Dover and Simcoe were flooded with airmen, many accompanied by wives and children seeking accommodations. Local newspapers reminded residents it was their patriotic duty to offer rooms wherever possible and to do whatever they could, "to offer friendly welcome to these men who are giving their services in one of the most important branches of the Empire's defence."[3] The response from townspeople, cottage owners and businesses was overwhelming. Even farmers, some already hard-pressed to provide adequate sleeping arrangements for members of their own family, were among the first to offer lodgings at reasonable rates. Everyone, it seemed, felt obligated to convert unused rooms into apartments wherever they could, many residents unselfishly disrupting their own households to do so.

1940 - War Comes to Walpole

One of several boats brought in from Trenton for use on the bombing range. The boats were used for placing splash targets and patrolling the danger zone marked off by buoys.

Back at the airfield, meanwhile, things were slowly improving. Amid great excitement, the first six Fairey Battle bombers touched down on partially completed runways, followed a few days later by a shipment of 87 octane aviation fuel which arrived at the Jarvis train station by tank car. To service the planes, aero engine and airframe mechanics reported from Technical Training School, along with armourers and other essential ground crew. Three RCAF marine craft intended for use on the bombing range docked along the east pier at Port Dover and the motor transport section received two Ford ambulances. Four "flights", or detachments, consisting of A maintenance, B bombing, C

S/L G.C. Butchart, Officer Commanding Ground Instruction School, examines the record of a student in his office with (standing, l-r) Flying Officers Johnston, Lanyon, Irvine and Wood. NAC C74479.

A proud moment at Jarvis. Members of Course 1 become the first air gunners to graduate in Canada under the BCATP, October 28, 1940. On hand to present wings for the occasion was Air Vice Marshal Lloyd Breadner, DSC. LONDON FREE PRESS COLLECTION/UNIVERSITY OF WESTERN ONTARIO ARCHIVES

RAF students from all parts of Great Britain made up the bulk of this class, Course 92 Air Bombers. More than 42,000 British students graduated from BCATP schools in Canada during World War II.

gunnery and D target towing, were hastily organized and Group Captain George E. Wait of Ottawa arrived to assume command. Thirty-nine pupil air observers reached the school on August 18 and a week later the first air gunners commenced training.

To fly the planes while the students trained, qualified pilots reported daily from the Central Flying School at Trenton, Ontario. Conditions were far from ideal, to say the least, and for weeks every landing was a crosswind landing, the pilots having use of only an east-west runway with a north-south prevailing wind. Mustering all their skills, Jarvis pilots fast became experts at plunking down five-ton bombers despite as much as 45 degrees drift! Until the control tower was completed, the crash tender served as a makeshift control centre, with the duty pilot directing all flying with little more than binoculars and a Very pistol.

Early entries in the station diary describe the problems with the weather:

Station very muddy this morning from heavy thunderstorm during the night. ... Apron in front of 'F' hangar serviceable, E'

and 'D' still drying. ...
Rain in morning and
worst mud yet. ...
Bombing flight first
programme cancelled,
rain and low ceiling.
... Gunnery Flight
2nd programme can-
celled, rain and low
ceiling. ... Bombing
range opened with a
bang at plus 30 sec-
onds on [delayed]
schedule. First bomb

*Fairey Battles on the tarmac with Hangars E and
F and cement gunnery backstop in the background.
Only the backstop remains today.* DND/PL2638

hit 25 yards from target. Everyone highly pleased. ... Flying
Officer Knickerbocker stands Battle aircraft 1654 on nose in
soft mud. ... Heavy rain all morning. ... Lightning caused minor
damage to 'C' Hangar. ... First shot fired at a drogue at 1340 hrs.
... Flying washed out at 1510 hrs. Rain.[4]

Despite slightly less than favourable flying conditions, the
first air observers finished training on September 29 and on
October 28 the first air gunners to graduate in Canada received
their wings on the windswept tarmac of the school. At a special
dinner given in their honour at the Hotel Governor Simcoe the
night before, G/C George Wait, commanding officer of the
school, expressed his regret at their leaving and had nothing
but praise for the graduates.

"With this class graduating, the school is over the hump
now," said G/C Wait with jubilation.

> We have had little grief since the school opened and I personally
> wish to thank you for your splendid cooperation in the face of
> all difficulties. You have shown a splendid spirit since you
> arrived and I am sorry to see you go. You boys are air gunners
> and it is on your accuracy as air gunners that the whole aircrew
> depends. It is a high trust which I know you will uphold. In
> conclusion I wish you all the very best of luck when you reach
> your new stations.[5]

1940 - War Comes to Walpole

It was a proud moment for the 58 Canadian members of the course who had their coveted AG badges pinned on their tunics at 10:30 the following morning by Air Vice Marshal Lloyd S. Breadner, DSC, chief of the air staff. Also receiving their wings and sergeant's stripes that day were members of the second class of air observers to graduate from the school. Speaking briefly, the air vice marshal addressed the men, pointing out the importance of stations such as Jarvis in providing essential training for air gunners and observers. Following the ceremony the air gunners left for the customary leave, some by chartered bus from the airport, others in canvas-covered RCAF troop trucks. It was a boisterous bunch that arrived in Jarvis on the way to the train station early that afternoon and their lusty cheers as they passed through the town left no doubt as to their enthusiasm. While the gunners would remain in Canada to help meet home defence requirements, all of the observers were destined to report within weeks to active squadrons in England. Sadly, just one in three would return to Canada.

November saw a steady stream of officers and other ranks reporting to the school and in no time the number of personnel on the station had exceeded 700—the total population of the

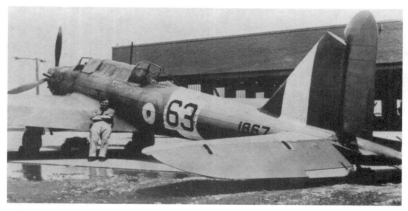

The Fairey Battle was exceptionally large for a single-engine airplane. The white painted band on the fuselage identified it as a gunnery flight aircraft. WALLY FYDENCHUK

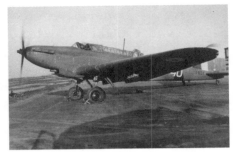

village of Jarvis. The majority consisted of airmen from all parts of Canada, with several Americans serving as staff pilots and instructors. With work on the barracks, mess halls and other buildings now proceeding faster than anticipated, an even greater number of

A ground crewman helps a pilot adjust his harness prior to a flight. WALLY FYDENCHUK

students could be channelled through the school—an unexpected development which made planners in Ottawa happy. Originally, plans called for the instruction of 130 pupils at any given time, a figure which would increase to over 400 during the next two years.

Unfortunately, things were not going quite as well with planes and equipment. Shipments arriving from England were irregular at best, forcing students to learn their skills on

New students are briefed on gunnery techniques using a plywood model of the bombing range. DND PL 4024

outdated machine guns such as the Vickers GO (gas operated) gun rather than the more effective .303 Browning. At one point even the delivery of Fairey Battle and Anson aircraft had to be temporarily suspended due to the critical war situation in Europe and

A student wireless air gunner takes aim with a Vickers K or VGO (gas-operated) machine gun. NAC PA185047

pressure on Britain to halt shipments to Canada. Fortunately cooler heads prevailed and the supply of aircraft and equipment resumed again. By early November there were 19 Fairey Battles in use at the school, shared equally by bombing, gunnery and target towing flights. To distinguish the aircraft in the air, a red band was painted around the fuselage just aft of the rear cockpit for bombing flight, while a white band identified those Battles belonging to gunnery. Target tug Battles were painted yellow with black stripes as a warning to other aircraft that they were trailing behind them a steel cable and 12-foot-long target sleeve or drogue.

The Fairey Battle had been designed as a light bomber in the late 1930s and was then considered well ahead of its time but by 1940 was declared obsolete and withdrawn from front-line duty. Indeed, so slow and defenceless were they that the first Victoria Crosses to be awarded the RAF in World War II were given posthumously to Battle crews who failed to return in alarming numbers from daylight raids in France. Dismantled, crated and sent to Canada, a total of 802 Fairey Battles would eventually see service with the RCAF. Although the planes were ruggedly built, crews at the training schools remembered them as smelly, oily and notorious for glycol leaks. Powered by a single 1030 h.p. Rolls Royce Merlin, the Battle became the mainstay of bombing and gunnery schools until replaced by more efficient Ansons, Bolingbrokes and Lysanders. While most Battles were painted yellow overall, many others still bore the camouflage paint scheme and markings of former RAF squadrons.

LAC Eric Cameron, a gunnery student in one of the first classes at Jarvis, logged all his training flights in the Fairey Battle. In the September 1993 edition of *Short Bursts*, the newsletter of the Ex-Air Gunners Association, he relates a harrowing experience he and fellow student Fred Brown had in one of the aging aircraft:

> One Battle staff pilot was a Yank who had visions of being an ace fighter pilot in the Battle of gunnery trainees. He tried to appease his frustration by indulging in some hair raising low flying and, occasionally, attempting aerobatics with the tired old Fairey Battles. One passion after the two gunners had taken their turn at the drogue—was to hunt for bridges he could fly under. One day, roaring along at water level up a narrow river, a very low bridge came into sight. Fred and I looked at one another in consternation because we were sure the Battle's wingspan was too wide to pass under that bridge. Closer and closer the Battle roared and we closed our eyes. Then the kite shot up at a steep angle and down again, almost ejecting us from the rear cockpit. "Fooled you fellas that time!" the pilot yelled and yanked the Battle into a tight turn—so tight the G-force pushed us to our knees.[6]

An even more dramatic incident occurred about the same time 6,000 feet above the frozen waters of Lake Erie. Two Toronto students, Howard Hewer and Frank Cook, were on a similar training flight in the rear cockpit of another Battle. When the pilot pulled up the nose of the plane sharply, Hewer was catapulted over the side. He was not wearing a parachute. Luckily, one of his flying boots became entangled in part of the equipment, preventing his tumbling all the way down into the lake. The pilot, oblivious to what was happening behind him and thinking the students were enjoying the ride as much as he, continued to weave back and forth. It took several seconds but Cook, a muscular type, eventually managed to clutch his friend by the pant legs of his flying suit and drag him back inside. It turned out that the safety wire (or G-string), normally attached to the rear floor of the Fairey Battle and designed to secure the gunner in place, was missing when the pair boarded for the exercise but they had neglected to report it. Upon landing, Hewer counted his blessings and made a point of informing ground crew about the missing life-saving wire.

AC 2 P.J. Deebank died trying to retrieve a drogue target. LOIS FOULDS

If Hewer was lucky, two other students were not. On November 8, Aircraftman 2nd Class Percy Deebank attempted to retrieve a drogue target which had become detached and fallen into the lake near Selkirk. Deebank, a popular 20-year-old Cornwall, Ontario native, was stationed at the school's 40-foot observation tower at Hoover's Point, and immediately set out to recover the target. Half a mile from shore his small boat capsized. He attempted to swim back but his heavy clothing dragged him under. Less than two weeks later LAC Joseph Boyd, a former McGill University student from Montreal, died instantly when he was struck by a whirling propeller on the main fairway directly in front of the hangars. According to reports, he had just returned from a flight in a Fairey Battle and was chatting with another student in front of the plane when he inadvertently stepped backward into the huge, three-bladed prop which was still turning upwards of 600 rpm. As a gesture of sympathy, his observer's wing was posthumously awarded to his father at graduation ceremonies on November 21. Boyd, who had attained third place in his class, was 19.

As the year came to an end, more American pilots arrived at the school. At the outbreak of war, Canada had a shortage of

LAC J.A.W. Boyd was killed by a
propeller on the school runway.
McGILL UNIVERSITY

trained pilots and many from the United States who flew with that country's airlines or US Army Air Corps were encouraged to enter Canada through the Clayton Knight Committee, later renamed the Canadian Aviation Bureau. This was an organization which allowed neutral American citizens to join the RCAF a full 18 months before the United States entered the war. Legislation passed by the Liberal government of Mackenzie King permitted Americans to enlist without having to swear allegiance to the King or to forfeit their US citizenship. An estimated 6,000 Americans eventually took advantage of this arrangement including some 900 experienced pilots.

Fifty-five of them were staff pilots at Jarvis in December 1940. One of these was Flying Officer Edwin Bounds, 32, from Vossburg, Mississippi. Arriving at the station on December 8, he was immediately assigned pilot duties with bombing flight. A large, heavyset man, Bounds had difficulty fastening his safety harness as none outfitted in the cockpit of the Fairey Battles was long enough to go around him. For this reason he never used it and directed ground crew members not to attempt to strap him in—an unwise decision that a year later would have tragic consequences.

Chapter Two
1941 - Setting the Standard

*T*he new year brought the managing director of American
Airlines from New York to discuss air traffic in the vicinity
of the air school. Back then American Airlines pilots were flying
daily between Buffalo and Detroit at precisely the same altitude
from which bombing and gunnery exercises were being carried
out over the lake. What's more, there was a growing concern
by the airline and air force alike about the temptation by RCAF
pilots to "jump" one of the big DC-3 passenger planes lumbering
along at 4,000 feet—a game which, while enticing, was strictly
forbidden.

By early 1941 the first stage of construction at the base was
completed. The building area alone covered almost 50 acres
and included a guardhouse with post office, headquarters, motor
transport section, officers' quarters, officers' mess, six airmen's
barrack blocks, airmen's mess, control tower, one hospital with
34 beds, drill hall, supply depot, two non-commissioned
officers' quarters, one NCOs' mess, civilians' quarters, civilians'
mess, library, works and buildings garage, 25-yard machine gun
range, ground instruction school, five bombing instruction
units, recreation building, large and small canteen, four indoor

Early view of the field looking south toward Lake Erie. Just out of view in the top right corner is the current site of Nanticoke power generating station. DND/REA-107-71

ranges for machine gun turrets, fire hall and dental clinic. Streets on the station were given such distinguished names as Rookie Road, Mackworth Mews, Raid Avenue and Sleepy Hollow. There was also a sports field, parade square, storage for 20,000 gallons of gasoline and a compass swinging base. The six large hangars each measured 160' x 244', included spacious lean-to offices and workshops along both sides and were lettered A to F. Hangars A and B were occupied by maintenance, while C was allocated to bombing, D to gunnery and E to drogue, with F hangar being used by the parachute section and as a spare. The four landing strips were 3,000 feet in length with a hard surface runway on each measuring 2,500 feet long and 150 feet wide. The 45,000 gallons of water required daily at the school were supplied by a pipeline from Lake Erie.

On February 17 the first 11 Norwegian students reported to the school. Norwegians were the first of the European Allies to arrive in Canada, many having been sent from England to train under the BCATP. Through arrangements made with the Canadian government, a pilot training facility known as

Little Norway was established at the Toronto Island Airport and later moved to Gravenhurst, Ontario. Those selected for other aircrew trades were posted to Canadian schools like Jarvis for training as observers and air gunners.

In March the school accepted the first of many airmen from Great Britain, Australia and New Zealand, as well as from the British colony of Newfoundland, the United States and Poland. Many of the Norwegians and Polish volunteers had been driven from their homes by the advancing German army and later made their way to England. Men from Britain had become embittered by the concentrated and devastating bombing of British cities and the destruction of Coventry in November 1940 made them anxious to complete their training and return to England as soon as possible. Although that country had its share of air training facilities, demands on airspace in the United Kingdom had greatly increased by this time, necessitating the transfer of RAF schools and students to Canada and the wide open space it offered. It proved to be a bittersweet experience for the young men from Britain. While they enjoyed the peace and security of a country far removed from the battle zone, their thoughts were never far from those they left behind. For the Australians in their distinctive dark blue uniforms (passionate purple, they called it) and the New Zealanders, the

Winter at the corner of Rotten Row and Pee Street. DND PL 4033

Sights on Jarvis

Leveling earth along runways and taxi strips continued well after they were in use. Here, local resident Mick Henning operates heavy equipment brought in by Johnson Brothers Contractors, Brantford. MICK HENNING

biggest challenge, perhaps, was adjusting to the harsh, cold temperatures and wintry blasts of southern Ontario. The majority of trainees were between 18 and 32, but some were even younger, several having falsified documents in order to enlist. Few possessed a driver's license.

As the first bombing and gunnery school in operation, Jarvis initiated and perfected many of the training procedures adopted at the other schools across the country. For this reason, the station received considerable publicity, particularly in the early stages. Forty-five newspaper reporters from across Canada and the U.S. visited the school early in 1941, followed a few weeks later by *Life* magazine photographers and a crew from Associated Screen News who filmed the training. Two officers with the US Navy paid a visit to the station in March to see firsthand the training methods used and Major C.G. Pearcy of the US Army Air Corps arrived by air to obtain information on the turret trainers.

In an effort to boost the recruitment of even more young men into the RCAF and to further promote the BCATP in Canada and the United States, Warner Brothers headed north in 1941. Touring several airfields, Norman Reilly Raine, a head producer with the studio, arrived at No. 1 Bombing and Gunnery School on March 25 in preparation for a full-length feature film starring Hollywood's James Cagney. The picture, with an impressive supporting cast that included the likes of Dennis Morgan, Brenda Marshall and Alan Hale, premiered a year later and featured RCAF stations Uplands, Trenton and Jarvis. A fictional story about bush pilots in northern Ontario who enlist in the Royal Canadian Air Force, *Captains of the*

Hollywood's James Cagney portrayed Jarvis staff pilot Brian Maclean in the 1942 motion picture Captains of the Clouds. DND/PL 5078

Clouds remains to this day a stirring reminder of Canada's war effort in the air.

June 1941 saw over 100 new trainees arrive at the school, including 63 wireless operator/air gunners reporting for duty from wireless schools in Montreal and Calgary. They included 57 Australians, five Canadians and an enthusiastic 21-year-old from New York City. The lone American in the group, Leading Aircraftman Duke Waddell, had enlisted in Canada in 1940 as a pilot but had been chosen instead by an RCAF selection review committee to take the gunner's course. He painfully remembered the evening he walked the entire five miles from the Jarvis train station to the airport in a pair of issue boots that were too small and the almost daily brawls that took place between the Canadians and large contingent of Australians on course. These scuffles, while rough at times, were "all in fun", he recalled with amusement from his home in California years later. Like the majority of Americans who enlisted in Canada early in the war, Waddell elected to remain in the RCAF rather than transfer to the US Army Air Corps when the opportunity arose later. He had, after all, benefited greatly from his period of training in Canada. Besides, he had no desire to leave a system he felt had been tried and proven.

Fairey Battles of bombing flight. At one point in 1941, up to 66 Battles were employed at the school. DND/PL 4028

On June 7, a special wings presentation in connection with the most recent Victory Loan campaign was held at No. 1 Manning Depot, Toronto for recent graduates of the bombing and gunnery schools at Fingal and Jarvis. Music was provided by RCAF station bands from Ottawa and Toronto. Wings were pinned on by Canada's highest-scoring ace of the First World War, Air Marshal W.A. "Billy" Bishop, VC, DSO, MC, DFC, who wore his own observer's wing for the occasion. Jarvis graduates consisted of 34 Canadians and three Norwegians, all members of Course 15, Air Observers.

Observers were one of five categories of aircrew trained at No. 1 Bombing and Gunnery School. Arriving from a 14-week course at an air observer school, these navigator/air bombers spent a further six weeks peppering Lake Erie shallows with 11½ pound practice bombs, pausing occasionally, as F/L Paul Burton once put it, "to allow the war to catch up."[1] Maps, charts, compasses, bombsights and cameras were among their tools of the trade. In addition, they were expected to send or receive eight words of Morse a minute and be able to man the machine guns if necessary. Upon completion of their course at bombing and gunnery school, students were sent to an air navigation school for advanced instruction in astro navigation, to better equip them for navigating by the stars.

At bombing and gunnery school, observers gained about 20 hours flying experience. From different heights and directions, students released a series of four, six or eight bombs on a 30-

Standard Department of National Defence floating bombing target, under construction. Targets were painted red. CWH/A. JUDD KENNEDY

foot triangular target painted red and anchored three-quarters of a mile from shore, ideally in a bay. Filled with titanium tetrachloride, these bombs gave off a puff of smoke upon hitting the water. The bomb hits were sighted from two quadrant shelters, observation posts built 90 degrees to one another on the bay. The accuracy of the bombing was phoned back to the airport and charted by the time the students landed. These telephoned reports contained the coordinates of the hits along with the exact second, minute and hour. To aid students in navigating, a large wooden arrow pointing to the target was positioned on the ground beside one of the two quadrant shelters.

Cpl. John Thompson of Vancouver sights a plume of smoke from one of the quadrant shelters on shore. Coordinates of each bomb strike were phoned back immediately to the plotting office at the airport. LONDON FREE PRESS

Most floating targets were constructed of timber and wood slats, shaped like a pyramid and kept afloat with the assistance of steel drums. After at least one of the floating targets was carried into the lake by receding ice that first spring, fixed wooden targets, secured on the lake

The makeshift shed used by the marine section based at the mouth of Nanticoke Creek. Most marine operations originated from Port Dover harbour. MILDRED ADDISON

bottom and filled with stone were also used. Targets were situated off Peacock Point (B.1) and Evans' Point (B.3), with one land target (B.5) located between the Third and Fourth Line of the Six Nations Indian Reserve, four miles north of Hagersville. Following expansion of the airfield in 1942, a fourth target (B.7) was placed in the marsh southwest of Turkey Point and a fifth (B.9) in the water east of Port Ryerse.

For the duration of the war, pleasure boats and commercial fishing tugs were restricted from entering the danger zone, a radius of 1000 yards from the centre of each target. Nevertheless, many locals took advantage of the opportunity to see bombing practice up close, from the relative safety of the shore.

LAC Harry Knipe (r) and off-duty Jack Paton in front of a quadrant shelter near Port Ryerse. Airmen assigned to huts were housed and fed in homes close to the range. JACK PATON

Packing a picnic lunch, entire families would drive to the bluffs overlooking the lake, watch each bomb strike the water then listen for the delayed bang. Such outings soon became a popular way to spend an afternoon or evening at a time when a ticket to a movie was a luxury few farmers could afford.

Funeral of LAC's Ray McNabb and Charles Taggart, Knox Presbyterian Church, Jarvis. The Australians were killed in a plane crash near St. Catharines on July 6, 1941. MYRTLE JOHNSON

At approximately four o'clock on the afternoon of Sunday, July 6, two Australian trainees, LAC's Ray McNabb and Charles Taggart, were killed when Fairey Battle 1803 crashed and burned at Decew Falls, three miles south of St. Catharines. They had just completed a camera gun exercise at 3,400 feet when dense volumes of smoke were seen emerging from the engine and pilot's cockpit. The crew of an accompanying plane saw the pilot, F/O Harold Moore, making frantic efforts to communicate with the pupils through his speaking tube. These efforts were continued down to 2,200 feet at which time the pilot was forced to bail out. The pupils never left the plane, which struck the embankment of a reservoir, flipped upside down and exploded. F/O Moore, in shock and burned about the face, landed in some trees on the bank of Twelve Mile Creek, still shouting for his crew to get out.

Owing to wartime conditions, the bodies of the two men could not be sent home for burial but instead were laid to rest with full military honours the next day at the Knox Presbyterian Church cemetery in Jarvis. Dozens from the village turned out to pay their respect as two RCAF ambulances, preceded by a military guard, carried the victims' remains from the Holmes Funeral Parlour along Main Street to the church. Completing the procession were 50 fellow Australian trainees and a long line of officers and airmen from the school. Inside the church, the flag-draped coffins were banked on all sides by floral tributes from the community while members of the three local churches made up the choir. Because of problems moving the coffins around the church railing, this would be the only RCAF funeral conducted from Knox. Future services were held either at St. Paul's Anglican in Jarvis or at the recreation hall on the station.

A court of inquiry convened for five days and concluded from evidence uncovered in the wreckage that one of the two students had inadvertently sprung his parachute while still inside the plane, rendering it useless. The other, the court reasoned, had already clipped on his own fully-functional chute, but remained behind to help his friend.

Happier times were in store a few weeks later when George Edward Alexander Edmund, Duke of Kent visited the school on August 27. Arriving by air, the youngest brother of King George VI and his entourage were given a royal tour by the commanding officer, G/C George Wait and senior officers, who explained the various aspects of training. An inspection of the control tower and C, D, and E flights followed, during which the duke took considerable time to stop and talk with aircrew and ground crew alike. At the station hospital he met and

HRH the Duke of Kent chats with members of the Royal Australian Air Force during his inspection tour of the base. MYRTLE JOHNSON

chatted with several patients, including LAC Jack Martin of New York.

The Duke of Kent, youngest brother of King George VI (l.) and G/C George E. Wait, commanding officer, tour the maintenance hangars on August 27, 1941. One year later, the duke would die tragically in a plane crash. TED POTTER

As he approached the student's bed, the duke asked what was wrong with the young man. Springing to his feet, Martin saluted and replied, "Only a sprained ankle, Your Royal Highness!" This tickled the duke into quite a distinct smile. It seemed like quite a mouthful when all Martin had to do was remain in bed and say, "Sir".

While witnessing bombing practice at the range that afternoon, the duke expressed a desire to take a ride on the 42-foot crash tender anchored a few hundred yards off Woods' Point. A dinghy was provided and together with G/C Wait and other members of the party, the duke boarded the cabin cruiser. They then proceeded to nearby Port Dover where dozens of residents and tourists, learning quickly of the impromptu visit, lined the pier. The duke was greeted with enthusiastic cheers as he stepped from the boat and made his way through the crowd to the car which was to take him back to the airport. Following dinner in the officers' mess that evening, all the officers on the station were presented to him. He remained overnight, departing by automobile for St. Thomas the next morning along with members of the royal party and Ontario premier Mitchell Hepburn.

August 1941 also saw the official opening of No.16 Service Flying Training School (SFTS) eight miles away at Hagersville. This brought to five the number of airfields operating in Haldimand County under the British Commonwealth Air Training Plan. Other RCAF stations included No. 6 Service

Flying Training School, 25 miles to the east at Dunnville, and a relief landing field complete with hangar, barracks, mess hall and motor transport garage near the village of Kohler. A second relief field, known as R.2 Dufferin, would later be established south of Hagersville to help alleviate congestion on the main aerodrome there.

It was at SFTS that new pilots were introduced to the next stage of flying. After spending about 60 hours learning to fly rugged primary trainers like the Fleet Finch at elementary flying training school, students reported to one of 29 service flying training schools in Canada. Initially, the RCAF divided each SFTS into one of two categories—one for bomber pilots, another for fighter. At Dunnville potential fighter pilots received instruction on highly manoeuvrable Harvards and Yales, while at Hagersville students learned to master the more cumbersome Anson. Course lengths varied from nine weeks in 1941 to 16 weeks in 1943 and five months in 1944.

In terms of flying time, these airfields were situated mere minutes from the bombing and gunnery school at Jarvis. On any given day, as many as 350 aircraft could conceivably be in the air at the same time, though remarkably few mid-air collisions occurred as a result. While most bombing and gunnery flights originating from Jarvis were confined to the shoreline of Lake Erie and the Niagara Peninsula, student pilots on cross-country exercises from Dunnville and Hagersville often ventured as far north as Midland and as far west as Sarnia. Occasionally, fledgling pilots even found themselves hopelessly lost south of the lake over Pennsylvania or New York State— a stroke of bad luck which often would see the unfortunate students "washed out".

By autumn the number of personnel at Jarvis had increased to almost 1,500. To help meet the social needs of so many, a YWCA Hostess House was opened. Located just inside the main gate, this building became a home away from home for hundreds of airmen and women. Here they could write home, read a book, play darts or take part in a singalong around the piano. It also provided a place for wives and families to gather for an evening or for folks to pass the time before being taken on a tour of the camp. Easily identified by its roadside sign bearing the YWCA triangle and slogan "A Friendly Place To Meet", the hostess

An all-yellow Bolingbroke on a high-level gunnery flight. The revolving mid-upper Bristol turret is clearly visible. NORM WILLIAMS

house quickly became one of the most popular spots on the base.

In October the first of several Bristol Bolingbrokes arrived from Yarmouth, Nova Scotia. The Canadian-built version of the RAF Blenheim medium bomber, the Boly had a wingspan

A Bolingbroke with minor damage to port cowling and leading edge of wing. This unique photo clearly shows the twin bomb racks installed on the belly of the aircraft for high-level bombing. JOHN C. MILNER

Sights on Jarvis

of 56 feet and was powered by either two 870 h.p. Bristol Mercury or US-supplied Pratt & Whitney SB4-G engines. With its revolving mid-upper turret, the Bolingbroke was ideal for machine gun practice and quickly joined the aging fleet of Fairey Battles serving gunnery flight. Eventually Bolingbrokes would participate in high-level bombing exercises as well, with practice bombs attached externally to racks beneath the fuselage.

Harvard trainers from Dunnville. Traffic was heavy at the service flying training schools with about 4,300 hours flown each month and 180 take-offs an hour. NO. 6 SFTS ASSOC.

Between 1939 and 1945 American Airlines maintained its pre-war routes across southern Ontario. To help passenger planes reach their destinations, two radio range stations had been placed strategically on the north shore of Lake Erie. One such post, operated directly by American Airlines since 1934, was located just a mile north of the Jarvis airport and it was from here that radio operators kept in touch with commercial airliners flying the Buffalo-Detroit route. Originating in New York City, these flights generally ended in Chicago with regularly scheduled service being offered both day and night. The radio range station remained in service at Jarvis until the spring of 1941 when it was amalgamated with the Strathburn range and moved to Clear Creek, just west of Long Point.

On the night of Thursday October 30, Captain David Cooper, pilot of American Airlines Flight 1 AM-7, radioed ground operators in Detroit and Chicago that they had passed above Jarvis at 9:39, altitude 4,000 feet and that everything was normal. But just half an hour later, the big DC-3 flagship plunged to the ground and burst into flames at Lawrence Station, Ontario, 14 miles west of St. Thomas. While the cause of the crash remains a mystery, it was widely believed at the time that the plane encountered a flock of wild geese migrating south and had taken one directly through the windshield. All 17 US passengers and a crew of three died that night in what was then Canada's worst aviation disaster. It seemed

The Jarvis train station and telegraph office circa 1940. Large groups of airmen came and went by rail and the bodies of Canadians and Americans killed at the school were also shipped home from here. BILL FESS COLLECTION

ironic, in view of the airline's ongoing fears of a possible collision between one of its DC-3's and some rambunctious military pilot, that the tragedy would occur instead on a drizzly night when the pilots of several RCAF flying schools along the route, including Jarvis, had all been grounded.

On November 18 a formal dance was held in the airmen's mess with music provided by the station orchestra. Throughout the war similar dances took place regularly in the sergeants' and officers' mess and eventually the much roomier recreation hall. While there was no shortage of servicemen, girls had to be brought in from the town of Simcoe and surrounding area as arranged by the local chapter of the Imperial Order Daughters of the Empire (IODE). Barely out of high school,

Looking west from the intersection of Highways 3 and 6 in the village of Jarvis. Many an airman hitchhiked from this very spot. Forty-eight hour leaves were frequently spent in Hamilton, Toronto or the U.S. BILL FESS COLLECTION

A station dance in the recreation hall. It was not uncommon for a thousand to attend.
NORM WILLIAMS

young ladies 18, 19 and 20 years of age arrived by the carload with the IODE graciously supplying the chaperones—one for every five girls.

Needless to say, the ladies were immensely popular. With up to 10 airmen for each girl, the evenings flew by all too quickly, though there was always time at the end of the night for the usual exchange of pictures, addresses and occasionally, a telephone number. Since a student airman's stay at the station in 1941 was limited to six weeks or less, dance partners rarely saw each other again unless it was at a wings parade or prearranged date on the beach at Port Dover. Many of the airmen continued to keep in touch long after they had graduated and were posted overseas.

In preparation for night bombing, pilots in bombing flight were given instruction in a Yale and several of them went solo. As all pilots were highly proficient, this amounted to little more than a refresher course. Most had logged thousands of hours flying time and virtually all had acquired night flying experience while employed with civil companies or during their military training. Still, there were increased risks associated

Social room at the officers' mess. TED POTTER

with night flying, not the least of which was being unable to see one another in the dark. To reduce the possibility of mid-air collisions, particularly over the target area, an elaborate system of two-way radio communication was worked out between the control tower and pilots. This meant pilots had to receive permission by radio before taxiing and again before flying over the target.

By late November night bombing exercises were underway at the station, the first ever planned for student air observers being trained under the BCATP. The same targets used during the day were used at night, illuminated so they were visible up to 10,000 feet under normal conditions. To provide electric power, submarine cables were laid from shore and lights on the targets controlled by personnel in the quadrant huts. On the main aerodrome itself, certain features incorporated during construction a year earlier now needed to be fully implemented. These included contact lights on each runway, a rotating beacon, illuminated wind tee and ceiling projector for measuring cloud height at night.

Bombing at night presented students with a number of problems which they did not have to face in the daytime; the most difficult of these was determining the direction and speed of the wind in the target area. This was partially overcome by checking the drift of the aircraft against fixed objects on the ground, such as headlands on the shoreline, which stood out clearly from the water. To help the students see where their bombs were going and to enable each bomb to be plotted at the quadrant huts, special flash bombs were used. These bombs gave off a bright flash of light lasting about one second—long enough for accurate plotting to be made in the dark.

Preliminary results of these exercises showed a high degree of accuracy, almost equal to that which could be expected in daylight.

To assist students in the air, classes were conducted at Ground Instruction School housed in building No. 2 at the station. The course of study for air observers, with its myriad formulas and applied mathematics, seemed like an awfully lot for an 18-year-old to absorb in just six weeks. Resembling the curriculum of a university course in engineering, studies included the theory of bombing, trajectory angles, meteorology, bombsights, wind drift and problems arising from it, fuses, carriers and use of flares. Since observers were also required to take a gunnery course, students had to learn the theory of sighting machine guns, tracer bullets, range estimation, gunnery tactics, types of turrets and the all-important art of aircraft recognition.

Following the unexpected attack on Pearl Harbor on December 7, 1941 and declaration of war by the US on Japan a day later, the hundred or so Americans at the school were left in a quandary. While most felt obligated to remain with the RCAF, others requested immediate transfers to branches of their own armed services. After turning in their air force blues, 23 staff pilots and 11 trainees at Jarvis eventually made their way to Toronto to resign from the RCAF in order that they might join the United States Army Air Corps. The days of Americans flocking to the border to enlist in Canadian services were over.

While returning to the airfield on Tuesday, December 9, two Fairey Battles clipped wings in the air just west of Fisherville. Two farmers, Harold Waterbury and Elvin Makey, were hunting rabbits when their attention was drawn to the planes approaching from the east. Looking up, they saw one aircraft roll over and two small objects fall from the plane moments before it slammed into the ground, still upside down. The second plane, though damaged, circled once before limping back to the airport to report the accident.

Authorities arrived to find LAC John Gray, a student air gunner, dead in the wreckage and two others unaccounted for. Following a hastily organized ground search by station medical officer F/L H.C. Moorehouse and his staff, two more bodies were discovered in a plowed field, half a mile away. "They were the objects we saw falling out of the plane," Waterbury recounted years later, "Though we didn't know it at the time."[2]

Dead were the pilot, F/O Ed Bounds, and LAC George Barber, a 31-year-old member of the RAF Volunteer Reserve

LAC George Barber with friend Enid Schneider of Guelph. Barber plunged to his death following a mid-air collision near Fisherville on December 9, 1941.
LOYS READWIN

from London, England. Bounds, identifiable only by his stocky build and custom-made blue flying suit, was wearing a parachute but had made no attempt to use it. Both men had fallen nearly 1700 feet.

The investigation which followed confirmed that Bounds was not strapped in and that the two students had been standing in the rear gunner's cockpit with the hood open. Only Gray had the safety wire attached to his harness, preventing his falling out. A court of inquiry recommended pilots be forced to wear safety belts.

On December 20 the first students from No. 4 Wireless School in Guelph to receive the single wing of an air gunner graduated at a wings presentation in F hangar. To mark the occasion, W/C Keith Russell, commanding officer of the Guelph school, was invited to pin on the wings and sergeant's stripes. These wireless operator/air gunners, or WAG's, received the same training as straight air gunners, but prior to

LAC J.S.W. Gray (l) and F/O Ed Bounds died after two airplanes clipped wings just west of Fisherville. NAC C14422/DND PL 2652

entering bombing and gunnery school were given an additional 20 weeks instruction on Morse code and use of the radio at one of four wireless schools in Canada. Academic requirement to enlist as a WAG was a minimum two years of high school.

In keeping with the highest traditions of the air force, G/C Wait and senior officers served Christmas dinner to about 600 enlisted men in the airmen's mess. This number represented those who were not on leave or spending Christmas with relatives or friends. Prior to December 25 every airman on the station was canvassed to see if he wanted to spend Christmas Day with a family in a private home. Several took advantage of the invitation and were bussed to homes in Jarvis, Simcoe and surrounding area. It was just another way for families to show their appreciation to the men in blue, many of whom were spending Christmas away from home for the first time.

As 1941 drew to a close, the station received word that P/O Laird Jennings, a graduate of No. 2 air observers course, had been awarded the Distinguished Flying Cross for conspicuous bravery and devotion to duty.

Chapter Three
1942 - The Terrible Summer

By 1942 the economy in Haldimand and neighbouring Norfolk County had received a tremendous boost. Stores, restaurants, theatres, banks and local churches all became busier places with the arrival of the air training schools and an army camp in Simcoe. To accommodate the growing number of servicemen, the Summer Garden dance pavilion in Port Dover remained open all winter and the ever-popular Arbor lunch counter was expanded to include a winterized dance floor featuring a Coke bar and jukebox. As airmen and their families continued to arrive, Port Dover's population alone grew by 25 per cent. To provide affordable transportation, a regularly scheduled bus service was begun which linked the town of Simcoe to the airport at Jarvis and later to Port Dover.

This increase in activity affected many people in different ways. While businesses flourished and merchants prospered for the first time since the Depression, households found themselves faced with critical shortages of such essentials as meat, sugar, coffee, butter and of course, gasoline. Wartime rationing, imposed early that same year, didn't help. Nor did the steady stream of air force families posted to the area whose purchases put an even greater strain on items already hard to

Sights on Jarvis

The fishing tug Stuart B *services the target at Peacock Point, January 15, 1942. Like other fixed targets in the lake, it was held in place with 600 tons of stone. Targets were liable to be in use day and night, Sundays and holidays included.* KEN PAINTER

come by. An accelerated training program proved a further annoyance to some. The introduction of night bombing meant many sleepless nights for residents as training planes rumbled overhead on their way to the targets—most nights, all night. Port Dover youngsters like Stan Morris and Jack Struthers, on the other hand, welcomed sunny days when they could cycle the seven or so miles along dusty roads to the airport. There they would clamber to the top of perimeter fences as aircraft took off and landed so low above them they could feel the slipstream.

In February, two months after the United States entered the war, a crew from the Columbia Broadcasting System arrived at Jarvis to broadcast a wings presentation on the CBS radio program, "Spirit of 1942." Special wires were laid at the station and the program carried by telephone to New York, where it was put on the air. The program, rebroadcast that same day over CFRB in Toronto, was enthusiastically received in both

The station band. THE FLY PAPER

the U.S. and Canada. It featured various phases in the training of observers and air gunners and included talks on armament, air gunnery and aircraft identification. Providing music was the RCAF band from No. 1 Manning Depot in Toronto. To conclude the program, G/C Wait broadcast his farewell message to the graduating classes as he pinned on their wings.

"You have come to the end of your course. It is with great pride and pleasure we present you with your wings and sergeant stripes," he told the graduates.

> With the exception of a short navigation course for the observers, your training is now completed in Canada. The decisive weapon of the air force is the bomb and it has been known for a long time that the issue of this war would be decided by bomber aircraft. That is why the Commonwealth Air Training Plan centered here in Canada has placed such stress on the training of bomber crews.[1]

Among the 34 observers to graduate was LAC C.F. Kirk of Toronto, who had his badge presented by his mother, Mrs. William Kirk. One of the most touching moments of the

An RAAF student tests his accuracy on the 25-yard range with a Vickers 'K' gun. Getting used to the noise was as much a challenge for some as firing the guns.
DUKE WADDELL

ceremony came when LAC Kirk leaned over and kissed her as she pinned on his wing.

On February 20, the 1000th air gunner to receive his AG wing at the school graduated from Course 26. At its peak, the station graduated double classes every two weeks, one class consisting of observers or air bombers, the other of air gunners or wireless operator/air gunners. The majority of graduates were promoted from leading aircraftman to sergeant and received an increase in pay to $3.60 per day. Like pilots, one-third of air observers received commissions as pilot officers. Few failed, though occasionally pupils were held over. Incurable airsickness, due in no small part to nauseating oil and glycol fumes in the old Fairey Battles, accounted for some of those who did not make the grade. A further four weeks training at air navigation school awaited the observers, while air gunners were posted to embarkation depot immediately following the customary three weeks leave.

1942 - The Terrible Summer

A Fairey Battle of drogue flight cruises west along the Lake Erie shoreline. Battles served as target tugs until replaced by Lysanders in March 1942.

The first experience students had with actual machine guns was on the 25-yard range located on the field just east of Hangar F. Here novice gunners took turns firing short one-second bursts at targets in front of the gunnery back-stop, 17 to 20 rounds at a time. A larger, more complex 200-yard machine gun range at Hoover's Point, nine miles to the east, provided students with even newer technology as they fired at targets from a hydraulically operated gun turret. The turret, which packed the punch of four .303 inch calibre Brownings, was of the Frazer Nash and Boulton-Paul types manufactured in Great Britain and shipped to Canada for use at training schools. The main purpose of the range was to give gunners and observers the experience of shooting from an actual tail turret using belt-fed guns capable of

A wireless air gunner prepares to fire on a drogue. Air-to-air firing with live ammunition was carried out in an area extending 18 miles from Peacock Point to Evans' Point. To enable a gunner to see where he was firing, tracer bullets were inserted approximately every three rounds. JOHN C. MILNER

Four of the planes used at Jarvis. Clockwise, from top left: Fairey Battle I, Bristol Bolingbroke IV, Westland Lysander III and Avro Anson Mk II.

firing 4,800 rounds per minute. All day long and into the night the powerful Brownings could be heard on the Point, their monotonous chatter spewing an endless stream of bullets out over the lake.

All aerial practice with machine guns took place in the restricted area along Lake Erie. To begin, students first fired tracer at the water to get an idea of curvature, drift and the feel of the thing. Next came splash targets, a row of wooden buoys moored in the water parallel to the shore, and finally drogues, fired at from different speeds and angles. Shooting at drogues was the closest experience to combat a potential air gunner received before drawing his wing and sergeant's stripes. Measuring just three feet by 12, the drogue billowed out much like an outstretched windsock, weaving in the breeze and proving elusive to many a gunner trying to knock holes in it from 200 yards away. Considering the field of fire, variation in turrets and gun mounts, instructors considered an 8 per cent score on the target as pretty good.

Coordination between gunnery and drogue flights was key. Every five minutes in flying weather, an aircraft would take off with two students aboard, the pilot rendezvousing with

Clearing snow-covered runways with a Sicard snow blower. Keeping runways open in winter was no small task. CHAA

the drogue plane at a specified time and altitude along the drogue line. This imaginary line stretched the entire 18-mile length of the bombing and gunnery range and it was from here that all air-to-air and air-to-ground practice with live tracer ammunition was done. Positioned in either the rear cockpit of a Fairey Battle or the mid-upper turret of a Bolingbroke, students opened fire on the drogue trailing several hundred feet behind the other aircraft. To reduce the number of training flights and for the tallying of individual scores, bullets were dipped in paint: those fired by one gunner left red-rimmed holes in the white fabric of the drogue, the other blue. The drogue plane would then circle a designated field where the drogue operator would release the bullet-riddled target and send a fresh one down the cable to take its place. When beginning a series of exercises, or when coming back to land, an electric winch installed in the back of the plane allowed the operator to reel the spool of cable in and out. Understandably, crewing a target tug could be an unpopular assignment, though there is not a single recorded case of a stray bullet ever striking a drogue plane at the school.

Another form of instruction air gunners received in the early stages of their training was the cine-camera gun. Loaded with 16 mm film instead of bullets, these guns could be aimed and fired at an accompanying aircraft which was simulating an

An Anson II. The Mk II was the Canadian-built version of the British MK I, but with American-supplied Jacobs L-6MB engines replacing Armstrong Siddeley Cheetahs.

attack. When processed, the film showed if the pupil was allowing enough deflection or if he was off the target. The advantage of these exercises were twofold. First, they spared the 200 to 300 rounds fired by each student per exercise, and secondly, camera gun practice could be carried out anywhere over land without the fear of spent ammunition coming down on populated areas. Until their course was extended in 1942, air gunners spent only about seven hours in the air.

On March 22 a freak snowstorm blew into the area without warning, catching several planes in the air. So severe was the sudden squall that one young pilot from Hagersville descended a short time later to find himself south of the lake over Dunkirk, N.Y. Two other students from the school, LAC's John Forst and James Clews, likewise became disoriented in the near-zero visibility, one breaking formation to climb above the storm, the other attempting to fly through it. While Clews successfully found his way to the airport at Kohler 25 minutes later, Forst flew into the ground near Selkirk with fatal results. Further east, five Dunnville Harvards running low on fuel were forced to land across the border, one at Niagara Falls, N.Y. and four more at the municipal airport in Williamsville. The following morning all pilots returned from the U.S. with bills for fuel and accommodations—and faces a shade more crimson than the station fire truck.

1942 - The Terrible Summer

Lysander 2426 against a Canadian winter landscape. With a bright yellow fuselage and broad black stripes, it was not surprising that Lysanders were affectionately dubbed "Bumble Bees" by some ground crew. DON SCRUTON

After serving over 19 months as commanding officer, G/C George Wait was posted in March to No. 1 Training Command as senior administration officer. A veteran of the Royal Flying Corps, Wait first joined the RCAF in 1922 and in the years that followed was attached to various air force stations and headquarters throughout Canada. One of the most popular and respected officers to have commanded in the RCAF, he became a great advocate of teamwork among members of aircrew and ground crew alike. According to graduates of No. 1 Bombing and Gunnery School, he was a living example of his advice. He was succeeded by W/C William Hannah of Winnipeg.

One of the first tasks the incoming CO had the pleasure of performing was to welcome and inspect the first 70 members of the RCAF Women's Division to be posted to Jarvis. Originally formed as the Canadian Women's Auxiliary Air Force, this branch of the service was later called the RCAF Women's Division. At Jarvis, some of the duties undertaken by these girls in uniform were that of cook, switchboard operator, fabric worker, photographer, clerk, parachute packer, canteen stewardess, accountant, stenographer and a host of other trades relating to administration.

A Bolingbroke in an unflattering position in the grass infield directly north of the hangars. Such occurrences were not uncommon. MICK HENNING

Corporal Ethel Davis, 22, of Edmonton was a member of that initial group of WD's. She recalled how they were received with some apprehension at first, but on the whole were generally accepted by the men. Within six months airwomen were serving in other areas of the base, including the control tower,

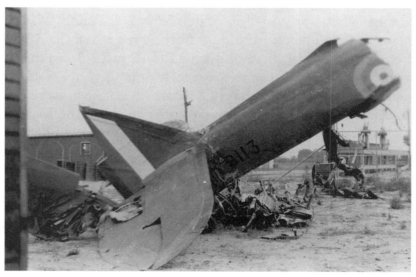

All that was salvaged of Bolingbroke 9113 after it crashed in Lake Erie, July 1942. Immediately following a crash all engine and airframe logs were impounded, as was the logbook of the pilot. CWH/ED HARSE

motor transport section, bomb plotting room, armament section, hospital and aircraft maintenance. Their eagerness to sign up meant that significant numbers of airmen previously confined to ground duties at the training schools were now free to enlist as pilots, gunners and other aircrew. By war's end more than 17,000 young women had enrolled as WD's including 1,450 posted overseas with the RCAF.

F/S Ken Norfolk in bomb aimer's position inside a Bolingbroke, July 1942. Early in the war crews communicated through a Gosport tube and headsets wired into their flying helmets. NAC C74463

By the spring of 1942 the number of aircraft on the station had surpassed 100. Included in this total were nearly 50 Avro Ansons and Westland Lysanders recently acquired by the school. Like the Fairey Battles, Ansons saw limited service early in the war until relegated to training and shipped to Canada. In addition to the 1,500 or so supplied by Great Britain, production of nearly 3,000 more was supervised by Federal Aircraft Limited of Montreal, earning the Anson the distinction of being the most widely used aircraft in the BCATP. Featuring built-in bomb bays, the Anson proved ideally suited for bombing practice while the Lysander assumed the role of target tug. With its high wing, large wheel spats and unmistakable black and yellow stripes, the Lysander became a familiar sight along the Lake Erie shoreline and at bombing and gunnery schools right across Canada.

For many of the staff pilots, chauffeuring students about in the air was largely uneventful and most yearned for overseas

Kitchen staff outside the other ranks mess. Monthly amounts of meat, eggs, fruits, vegetables and baked goods required to feed the base was staggering. ETHEL McNEILLY

postings. To be selected as a staff pilot or instructor in the early, crucial days of the war was often a credit to a pilot's ability, though few viewed it as such. Some, like P/O Bill Olmstead, considered a pilot posting to bombing and gunnery school as punishment for the unruly. Just weeks before he arrived at Jarvis, Olmstead had torn the instructor's certificate from his logbook, destroyed it and dropped the pieces on the chief flying instructor's desk—all because he had been denied a request for an overseas posting with fighter command!

To alleviate boredom, the more adventurous resorted to unauthorized low flying and sometimes they got caught. On June 24 a sergeant pilot from Jarvis buzzed the city of Hamilton and disappeared into the haze, but not before several conscientious citizens jotted down the aircraft serial number. It proved to be a costly bit of sport for the young flyer, who was later found guilty, reduced to the ranks and ordered to undergo 60 days detention. Similar escapades over the streets of Port Dover led to further threats of court martial. In a letter

A WD stores belts of .303 ammunition for use in the Brownings. WD's played a vital part at RCAF training stations. DND PL 12768

addressed to Port Dover village council, G/C V.H. Patriarche, officer commanding No. 6 SFTS, Dunnville, stated, "It is very desirable from the standpoint of safety of civilian life and property as well as for the safety of young pilots who are sometimes inclined to endanger their own lives without appreciating it, that any incidents of dangerous low flying be reported."[2] In the same letter, residents were urged to take down the number of any plane putting on a show and to phone it in immediately to the airport. Flying was strictly no-nonsense and RCAF authorities came down hard on those found guilty of bending the rules.

A series of accidents plagued the school beginning in June 1942. In just two months, four Bolingbrokes crashed on take-off, three Fairey Battles crashed on landing, two Lysanders ground looped in crosswinds and an Anson was forced down near Dunnville. Accidents were categorized as A, B, C or D, depending on the extent of damage to an aircraft. A D-category

LAC S.S. Schwartz (l) and P/O K. Slater. Schwartz, an American, and Slater, a member of the RAAF, died when their Bolingbroke Mk IV crashed into Lake Erie near Peacock Point. NAC C144234/RAAF

crash was of a minor nature, a C was usually repairable on the station, a B resulted in repairs having to be carried out at a repair depot, while an A-category was a write-off.

Word of a crash spread quickly. Farmers and housewives rushed to neighbours' homes with news or relayed what information they knew over the phone. It was not uncommon for the airport to receive several calls from distraught residents, though directions to an accident were often vague at best, resulting in ambulances and crash tenders occasionally rushing to a wrong location. Once at the scene, RCAF guards armed with rifles quickly took charge. For security reasons pictures were forbidden and anyone caught taking snapshots had cameras confiscated or the film destroyed. Keeping the curious from getting too close was another problem. In one instance, a young woman approaching a wrecked plane ignored repeated warnings to halt, until one determined guard drew back the bolt of his Ross rifle and raised it to his shoulder. The girl quickly retreated, but not before gazing back at the stern-faced airman still clutching his rifle. The two would see each other again many times after that and by war's end would be married.

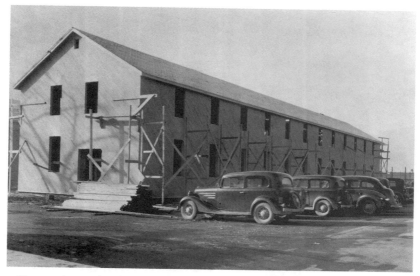

Construction at the airport was ongoing. Here a two-storey barrack block is added during a $250,000 expansion of the school in 1942. MICK HENNING

Fifteen-year-old Charles Cox lived just three miles from the airport and was typical of most youths in the area. When a plane was forced down, he would jump on his bicycle and be among the first at the scene to assess the damage, chat with the crew and perhaps collect a souvenir or two. Minor accidents were plentiful so close to the station and young Cox made a point of being at each one, regardless of where it happened to be. On July 23, he didn't have to go far.

While he was pitching straw on the family farm a short distance from No. 1 target at Peacock Point, he heard the blare of engines high above the bay. Glancing south, he caught a brief glimpse of a large plane in a steep dive, quickly gaining speed before it disappeared behind some trees. Charles rushed to the water's edge, where he spotted an RCAF crash boat speeding into the lake, where a Lysander was circling an oil slick a mile from shore.

The downed plane, a Bolingbroke Mk IV piloted by P/O Kenneth Slater of Australia, lay in 17 feet of water. Slater's body, along with those of LAC's John Williams, RAF, and Solwyn Schwartz, an American, were recovered from the

Sights on Jarvis

A later photo of the school that includes all the buildings following the 1942 expansion. The airport covered nearly 600 acres. STEVE MOUNCEY

submerged wreckage by divers the next day and removed to the Ivan Holmes Funeral Parlour in Jarvis. Schwartz, who hailed from Port Chester, New York, had served for a while as a second lieutenant in the US Marine Corps and had been in Canada for nine months. The night before his death he wrote a lengthy letter to his wife Jeanne in Toronto, telling her of his narrow escape the previous day. He explained how, after completing their bombing practice, the pilot had returned to the field and was about to land when he discovered one wheel jammed in the up position.

> You can imagine how I felt. We were in touch with the airport by radio and talking back and forth with them. We could see the fire engine and the ambulance on the field waiting for us to come in. There was a big crowd on the field and the whole thing made me feel as though I were in a moving picture. It didn't seem quite real. We dived and turned and shook the plane and at last the wheel came down and we were able to land perfectly. I was never so glad to feel the good old earth again.[3]

Ten-year-old Larry Hare attempts to drive the tractor straight while eyeing a Bolingbroke on final approach. LARRY HARE

It wasn't until 12 days after the crash that the body of a fourth occupant, LAC Harold Burnep, an aero engine mechanic from Toronto and father of two, was discovered in the water by fishermen 18 miles east of Nanticoke. Though the possibility of engine failure could not be ruled out, the primary cause was believed to have been an error on the part of the pilot, who permitted his airspeed to drop to stalling speed while allowing the drogue plane to overtake him.

A fatal crash cast a dark gloom over a station. There was the sense of loss experienced by roommates and fellow course-members, and the sobering realization it could happen to them. There were the military funerals with flag-draped caskets, followed by the traditional rifle volley and bugler's sounding of last post at the gravesides. When a victim's remains were shipped home, an officer or airman—frequently a friend of the deceased—always escorted the sealed casket. Though details of the accident were seldom made public, information arising from the mandatory court of inquiry inevitably found its way to the offices of flight commanders, or to maintenance squadron if the cause was found to be mechanical. In addition to the initial telegram sent to next-of-kin, a more detailed letter of condolence from air force headquarters was forwarded to the family, followed by a package containing the personal effects of the deceased. This often included the victim's logbook, identity discs, cap, wings and other badges he might have earned. As a further gesture of sympathy, wives and mothers each were sent a Memorial Cross (or Mother's Cross as it was more commonly known), a small medal on which the airman's name, trade and date of death was inscribed.

Blurred by speed in this photo, an Anson takes off into a southeast wind just above the Hare family home. Flying continued both day and night.
LARRY HARE

In August the first class of special air observers taking the moving target course finished their training at the school. These observers from No. 1 Central Navigation School took specialized training, bombing armoured motorboats that cruised up and down the shoreline.

Norman Davidson of Toronto was a member of that elite group. He was back at Jarvis a second time, having previously taken the standard six-week observer's course at the school earlier that year. Following an advanced navigation course at Rivers, Manitoba, he returned to Jarvis on July 20, 1942, where he spent the next three weeks taking aim at the moving target boats from Bolingbrokes flying at high altitudes. His logbook records eight entries during this period, commenting on everything from rough water forcing one boat to return to port to his displeasure at the wind information provided by the meteorology section. An amusing single entry in his logbook for August 5 reads simply, "Best exercise yet—1 bomb dropped within a few yards of boat—boys in boat worried."[4]

While the fast moving boats were seldom struck, there were exceptions. On one occasion, a student scored a direct hit on practice boat M.498, seriously damaging it and leaving one member of the crew slightly injured. The bomb, it turned out, had dropped straight through a narrow, slotted opening in the top of the boat and detonated inside rather than glancing harmlessly off the armoured plating. Obviously, the student was either very good or very lucky but regardless, must surely have headed his class at the end of the day!

An air bomber peers intently through his Mk IX bombsight. NAC PA 136264

About this same time, the first air bombers to undergo training in Canada arrived at Jarvis. This was a new category of aircrew that, together with the navigator, had begun to replace the air observer. Air bombers and air navigators B (navigator/air bombers) performed essentially the same duties as the observer but took a much longer course of training, which was eventually extended to 12 weeks. This provided double the amount of instruction originally given to observers in the first two years of the war and was deemed necessary to improve the overall accuracy of bombing operations overseas. To make room for the growing number of students reporting to the school, construction was started on several more barrack blocks. An additional north-south runway was also added.

For those living beside the airport, life was never dull. What once had been a quiet country existence had been transformed overnight into a hectic, military atmosphere. No one knew this better, perhaps, than Russell Hare and his family. Living just half a mile away from the airport, Russ, his wife Helen and four young boys farmed

New Zealand students receive instruction on the bomb release mechanism found in British bombers. DND PL 3569

Forty-eight members of the security guard and RCAF Service Police at No. 1 Bombing and Gunnery School.

land just east of the station, a location which placed their house smack-dab on the flight path of the east-west runway. Three of the boys, Larry, Vic and Robert, were born prior to the war and have vivid memories of what life was like growing up so close to the action.

Vic, who was just seven years old in 1942, remembers many things but mostly the fun of chasing stray drogues that fluttered down and of his attempts to snatch them before the air force arrived. The field where the drogues were released ran along the concession road just north of the runways and it was here where they were collected and the coloured bullet holes in the fabric stamped and counted on long tables in the kitchen of an abandoned farmhouse. In the course of any given week, these targets often fell short of the designated release point and were fair game as far as the boys were concerned. On one occasion, Vic remembers climbing a tall elm tree with five of his friends to retrieve a drogue stuck in the top. They had just wiggled back down onto the ground when they were greeted by a young airman who appeared out of nowhere, demanded the drogue and thanked them for saving him the trouble! He rewarded them with a nickel. Another time they won the race to a drogue which had fallen into a hay field, rolled it up and hid it in a culvert until the airmen who had come searching for it left, frustrated. The silky-smooth nylon drogues weren't just considered war souvenirs among the locals. They served many practical uses on the farm and were turned into hammocks, sheets and pillowcases.

"It was harmless things kids did back then," Vic recalled from his home in Selkirk many years later. But on at least one occasion the fun was not as harmless as it seemed. Finding the remnants of a broken mirror one day, Vic, his brother Larry and a couple of friends positioned themselves strategically in a neighbour's field at the end of the runway and tilted the shattered fragments towards the oncoming planes. It took skill to catch the planes in motion, aligning the mirror just right to cast a patch of sunlight directly onto the swift moving twin-engine bombers.

"It was great fun catching those things in the light of the mirror," laughed Vic. Fun, yes, but not amusing to those in the control tower. Just minutes after they had begun, a patrol car was dispatched from the base with two members of the RCAF Service Police, (SP) who located the unsuspecting pranksters in the grass and explained that the reflected light was shining directly into the pilots' eyes! Of course the boys hadn't thought about that. Nor had they bargained for the tongue-lashing that followed. With the threat of jail looming over them if they ever tried it again, the boys had their precious pieces of mirror confiscated and were severely reprimanded not only by the service police but also by their fathers, who were provided with a detailed account by the annoyed SP's.

Larry, the oldest of the Hare boys, remembers his part in the mischief. He also remembers the many young boarders his parents befriended from all parts of Canada, mostly RCAF students who were married and whose wives had accompanied them to the school. At times there were as many as three couples living in the house, particularly during the summer months. Since there were only three bedrooms upstairs there wasn't room for any of the boys so they were forced to sleep all summer in a tent in the front yard. Fortunately for them, Mom and Dad accepted fewer boarders in the winter.

There were other memories too. For six-year-old Robert it was the thrill of watching the big, yellow Ansons barely clear the roof of the house. Sometimes they were so close he could see the wind in the trees and wondered if they would make it. He remembers too the relentless rat-a-tat-tat of machine gun fire on the nearby 25-yard range. It was noisy, to be sure, but at the same time exciting for three impressionable young boys lucky enough to have grown up next to an air force station during the war years.

Assuming command of the school from W/C Hannah that summer was G/C Allan Duncan "A.D." Bell-Irving, MC. A former pilot with the Royal Flying Corps, Bell-Irving had been wounded twice in aerial combat, first when facing the German ace Max Immelman and again by Manfred von Richthofen in 1916. Following the war, he returned to Vancouver where he formed the Aero Club of British Columbia and served as a wing commander in the RCAF auxiliary. He joined the staff of Western Air Command the day war was declared and later

One of the station ambulances, doubling as a hearse, leaves Knox Presbyterian Church cemetery following a funeral in the summer of 1942. HOWARD ELLIOTT

commanded No. 4 Service Flying Training School at Saskatoon. He was posted to the Headquarters of No. 1 Training Command, Toronto as senior organization officer in August 1941, a position he held until his arrival at Jarvis.

The station was still reeling from the crash of July 23 when tragedy struck again, this time near South Cayuga. On Tuesday, August 18, an Anson on low-level bombing practice collided with a Fairey Battle on a routine gunnery exercise at 2,500

Sgt. John William "Will" Whitehead, one of six killed in the mid-air collision of August 18, 1942.
MARY MILLER

feet above the school's B.3 target at Evans' Point. The Anson, throttles wide open, flew out of control into the yard of Park Austin and burst into flames while his daughter Geraldine watched in horror from the side of the house. The second plane, Fairey Battle 1604, dove into the lake 200 yards off shore and was completely demolished. Ironically, it was the same aircraft which had survived the fatal mid-air collision near Fisherville the previous December.

Killed were the pilots, Sgt. John William Whitehead of Kaeo, New Zealand and F/O Maxwell Poole of Winnipeg, along with students John Lefurgey, Gordon Burke, Archibald Reed and Warren Kirkby, all from Canada.

The charred remains of three of the victims were recovered from the burnt-out Anson that same night, while those in the Battle could not be removed until the wreckage was raised from

LAC A.C. Reed and LAC G.C. Burke were among six killed when two planes collided near South Cayuga. NAC C144232/NAC C144221

10 feet of water the following afternoon. As a result of this latest accident, F/O James MacKelvie of No. 1 Training Command ordered a reorganization of the station's bombing range, directing:

> Unit to investigate feasibility of controlling traffic of low level bombing aircraft at 2,000 feet over No. 3 target and gunnery aircraft letting down from 3,000 to 1,500 feet at end of drogue line and that flight paths of above mentioned aircraft do not cross in vicinity of No. 2 control tower.[5]

Following a joint funeral in St. Paul's Anglican Church two days later, the body of Sgt. Whitehead was laid to rest with fully military honours in the RCAF plot at Jarvis, while the remains of the five others were shipped by train to their homes in Manitoba, Ontario, Quebec and B.C.

The following tribute to two of those killed appeared in the station newspaper, *The Fly Paper* on August 20, 1942, just nine days before they would have graduated:

LAC George Winfield, 19, with bride Lenore. Eight weeks after the wedding Winfield was killed when the Lysander in which he was a passenger crashed and burned. ISOBEL JOHNSON

In Remembrance

LAC Burke and LAC Lefurgey, two of the first Canadian air bombers to undergo training in Canada, have courageously, and without flinching in the line of duty, willingly given their lives in the great cause of liberty. In so doing, they have left behind them many bereaved relatives and friends.

We of Course 58 feel very deeply grieved that these two young airmen have passed away from our midst and although our hearts are laden with great sorrow, we must and will, carry on and complete the task they commenced in order that their training shall not be in vain.

During the performance of our duties in the air force we will always be reminded that their presence is evident amongst us. We shall always be conscious of their joyous expressions, the echo of their laughter in our ears, and knowing that memories of them linger with us, we shall keep performing our daily task until our efforts and the efforts of others, have ended this inevitable fray.

Remembered by Course 58 Air Bombers[6]

The summer of '42 ended as tragically as it began, three weeks later. On September 9 a Bolingbroke flown by 24-year-old Sgt. Robert McCrank, in company with pupils Fred Hawke and Ron Killick, took off on a routine gunnery flight arriving over Peacock Point just after 4 o'clock in the afternoon. Spotting the awaiting target plane at 3,000 feet, McCrank proceeded east down the drogue line, flying parallel to the second plane with one student engaging in air firing practice. At Evans' Point the exercise called for the gunners to switch position and for both aircraft to circle and descend to 1500 feet before heading west back to Peacock Point.

As the aircraft approached Evans' Point, or roughly halfway through the exercise, McCrank's plane suddenly and inexplicably broke formation and flew inland away from the drogue plane. Sgt. Roderick O'Hara, pilot of the accompanying plane and a friend of McCrank, immediately took up chase although the Lysander he was flying was slower than the twin-engine Bolingbroke, which appeared to be going full out. He continued his pursuit for several minutes at an altitude of 2,500 feet

Sgt. F.J. Hawke. ELGIN COUNTY MILITARY MUSEUM

when he observed the bomber in front of him suddenly nose down and dive almost vertically into the ground. Clanbrassil farmer George Jepson and neighbour Margaret Roberts heard the plane approach from the south then saw it plunge at terrific speed into a wheat stubble field and disintegrate. All three occupants died instantly.

Funeral of LAC George Winfield, St. Paul's Anglican Church, Jarvis. ISOBEL JOHNSON

LAC R.W.G. Killick (l) and Sgt. R.N. McCrank. Along with Sgt. Hawke (opposite page), the three died in a mysterious crash in September 1942. NAC C144228/ LILLIAN McCRANK

"I can still see that big plane coming down and hitting," recalled Margaret with a shudder 55 years later. "Like it was yesterday."[7]

Just what happened in the cockpit of the Bolingbroke that summer afternoon will forever remain a mystery. A court of inquiry failed to produce any reasonable explanation for the strange behaviour of the plane or its pilot. Owing to the total destruction of the aircraft, it was impossible for RCAF personnel to determine definitely if structural failure or control problems contributed in any way to the accident.

On September 26 a unique wings ceremony took place on the parade square when 25 airmen, all formerly employed in general duties with the RCAF, graduated as air gunners. "There is a special interest on this occasion because this is the first complete course of re-mustered members of ground crew to get wings of aircrew here," G/C A.D. Bell-Irving told members of the graduating class. Unlike previous students, many of those receiving their wings had already attained the rank of corporal or higher, having earlier served as clerks, service police and security guards. One member of the class, Sgt. D.S. Storey, had accumulated over 772 hours as a drogue operator at Jarvis.

Cpl. John Milner (centre) discusses night-time maintenance schedules with two mechanics of A Hangar, November 1942. JOHN C. MILNER

"We full appreciate the value of your service in the RCAF and the spirit that makes you want to go into the air and take a combatant part," the commanding officer continued. "We congratulate you and wish you the best of luck."[8]

Leading the class was Cpl. W.R. Clow, 21, of Belleville, Ontario, who had served at the RCAF recruiting centre in Hamilton for two years. In addition to receiving his AG wing, Clow was presented with a gift from his classmates as honour student, received a commission as pilot officer and was married that same afternoon to Edith Hastie of Hamilton.

It was a close call for a pilot and his students two weeks later when the port propeller of an Anson broke off its hub at 7,000 feet and sliced through the nose of the plane, striking a student in the bomb aimer's compartment and breaking his left arm. As soon as the pilot, Sgt. G.O. Scott, was able to regain control of the aircraft, he instructed both students to bail out. Despite his injury, LAC D.A. Dunlop, along with fellow pupil LAC R.K. Moyes, was able to leave the plane successfully and land safely a short distance from the field. The pilot then headed back to the station, with the port engine barely attached, where he somehow managed a wheels-up landing. Not since the fatal crash near St. Catharines 15 months earlier had it been necessary for Jarvis flyers to hit the silk. By war's

Preparing for Christmas dinner in the airmen's mess, 1942. Though given the opportunity to spend Christmas Day with local families, many chose to remain on the station. IRENE MILLER

end, seven others at the school would likewise be forced to take to their parachutes, all without serious injury.

While night flying in the early morning hours of October 27, several pilots alerted the control tower to a fire burning out of control on the Wabash Railway line a few miles northeast of the station. Aware of the potential danger, a volunteer party of pilots and ground crew hastily organized and made its way to Nelles Corners where the locomotive of a New York Central freight train was ablaze and on the verge of exploding. Acting quickly, F/O Jack Williams and Sgt. Ray Picard secured an axe and crowbar from a nearby farmhouse and with the help of both local and air force volunteers, uncoupled the cars yet untouched by the blaze. One by one they rolled nine tankers and four freight cars containing flammable material a safe distance down the track. As they approached the last car however, the civilian volunteers refused to go any closer for fear of an explosion. With the help of just seven others from the station, the two pilots then managed to push the final oil tanker clear of the burning engine. In all, it was estimated that hundreds of tons of gasoline and other vital war materials were saved and the possibility of a chain reaction explosion averted. For their part in leading the others, F/O Williams was awarded

Sights on Jarvis

On special occasions and during inspections, the whole station was required to turn out on parade. Here, the photographer captured a portion of the school's complement of 1800 men and women.

the George Medal and Sergeant Pilots Picard, Mayhew and Tunstall the British Empire Medal. Five more received letters of appreciation from the chief of the air staff.

In a unique effort to promote the sale of vital war bonds, a large number of planes from Jarvis and Hagersville descended on Norfolk County on Saturday, October 31 in what would be the most spectacular air show ever seen in the area. The aircraft, flying in formation in two waves, converged on Port Dover at 3:30 that afternoon before making their way to the communities of Port Rowan, Delhi, Waterford and Simcoe. As the planes passed over each town, hundreds of leaflets were dropped, outlining Canada's third Victory Loan campaign and explaining the need for residents to pledge a portion of their earnings to purchase these all-important bonds. Every year RCAF personnel from all three local flying schools actively participated in these drives, the airmen and women of No. 1 B&G alone easily exceeding their objective of $72,000 for the purchase of bombs in 1942. Such was the spirit at Jarvis.

On November 9 the station suffered its fourth fatal crash in five months when Lysander 2325 crashed and burned three miles northwest of Hagersville on the Six Nations Indian Reserve. According to several eyewitnesses, the aircraft circled low over the area before diving out of view behind some trees bordering the main hydro-electric power line that stretched across the reserve. Joseph Hill, a Tuscarora Township farmer, was among those attending an inquiry two days later, where he gave the following testimony:

I was plowing in a field on the west side of the barn at about four o'clock on the afternoon of November 9, 1942 when I heard a loud crash behind me. I looked around and saw a burning aeroplane careening across the ground. It came to rest in a scattered pile of burning wreckage. I tied up my horses as quickly as I could and ran over to the crash. I found two bodies with burning clothing and I did what I could to put out the fire. It was too late however as they were already dead.[9]

Evidence gathered at the scene pointed strongly to yet another case of possible unauthorized low flying. Deep gouges left in the ground by the plane's wheels directly below the high tension lines led to speculation the pilot was either attempting to fly beneath them or was merely diving close to the ground and did not see the wires until it was too late. Killed instantly in the crash were Sgt. Pilot Norman Wade of Hapton Burnley, Lancashire, England and LAC George Winfield, a 19-year-old aero engine mechanic from nearby Port Rowan.

Mechanics were often required to fly. Following repairs to an unserviceable aircraft, it was standard RCAF policy for at least one member of the maintenance crew to accompany the pilot on a test flight. These flights enabled mechanics to check firsthand if the engines were performing normally and to verify that the problem had indeed been corrected. LAC Harvey Ohland of Fort St. John, British Columbia, was an aero engine mechanic at Jarvis and well remembers taking several such flights. For his trouble, he pocketed an additional 75 cents flying pay—a welcome bonus considering he earned only $2.75 a day.

The day-to-day servicing of planes at the school was assigned to individual flights, while engine changes and major repairs to the airframe were carried out by the men of maintenance flight, located in hangars A and B. Each morning brought a round of daily inspections on all serviceable aircraft as they were towed from bombing, gunnery and drogue hangars and parked in a single row on the tarmac. Once on the line,

Aero engine mechanics check over the Rolls Royce Merlin of a Fairey Battle. Other aircraft maintenance trades included electricians, instrument and airframe mechanics.

mechanics started the engines, checked oil pressure and magnetos, then shut them down again, a procedure lasting about 20 minutes. It was the further responsibility of the mechanics to ensure that aircraft were refuelled as required and to return to the flight line with pilots and students to man fire extinguishers and remove wheel chocks before taxiing. At night, and on days when flying was scrubbed, it was work as usual in the hangar as engines were gone over, oil levels checked and aircraft washed down. At the same time, mechanics known as riggers pored over the airframe, inspecting such things as the rudder, elevators, undercarriage, brake lines, control cables and so on. Following each 40 hours logged, a thorough check was given the engine, airframe and propeller, with a major inspection after 240 hours. The majority of tradesmen making up maintenance were graduates of the Technical Training School at St. Thomas or the Galt Aircraft School.

During the course of the war, the maintenance section at Jarvis took pride in achieving one of the highest serviceability

records of any station in Canada, with up to 90 per cent of its aircraft on the line and ready to go at all times. Still, not everybody was happy. Cpl. Bruce Garber, a 27-year-old aero engine mechanic from Fisherville, was grateful to be posted just eight miles from home but was not content with his duties. He would have preferred to have been assigned to the main maintenance hangars A or B but instead found himself doing mostly minor repairs in gunnery flight, washing aircraft and constantly towing them in and out of the hangar. After weeks of complaining, he was finally brought before the commanding officer, G/C Bell-Irving.

"Garber", the CO began, "I understand you're not happy. What seems to be the problem?"

Station firefighting services. Top of page, a standard pumper with firefighters and four-legged mascot. Above, RCAF crash tender No. 774. These Marmon-Herrington Ford V-8 trucks drove on all six wheels, enabling them to negotiate hills, snow, mud and ditches. Two crash trucks were on standby at all times, one to respond in event of an accident, the other to remain on base until all other aircraft had been recalled.
BETTY SCHOLFIELD

Air Observer Bruce Murray, 20, of Fort Frances, Ontario, steps from the rear cockpit of a Fairey Battle following a flight over nearby Lake Erie.
DND PL 3574

"Well, sir, I'm a mechanic but I'm not doing what I'm trained to do. I want to spend more time on engines."

"Garber", the CO lectured, "they tell me your farm is just down the road. Do you have any idea what the other men on this station would give to be posted so close to home?"

"I fully realize that, sir, but I still want to work on engines."

"Very well, Garber," Bell-Irving conceded. "Leave it with me."

A few days later Garber was informed he would indeed be spending more time in maintenance, as requested, and at the same time received an unexpected new posting—to Goose Bay! He didn't get home again for three years.

Chapter Four
1943 - In Full Swing

The station celebrated New Year's Day 1943 with dinner
served to all personnel in the airmen's mess, followed by a
dance featuring Mart Kenney and his orchestra in the recreation
hall. Highlighting the day's festivities was the awarding of the
Air Force Cross to F/L Lawrence Montigny, an American staff
pilot from Cleveland, Ohio in recognition of outstanding
performance of his duties in connection with flying training at
the school.

While taking off for a test flight on Sunday, January 3,
Bolingbroke 9964 lost power and crashed a mile east of No. 3
runway. The aircraft, which was completely destroyed, had
just climbed to 500 feet when the starboard engine cut out and
the plane plunged to the ground following a violent right hand-
turn. A mechanic, Cpl. William Doan of London, Ontario,
died instantly while the pilot, F/S George Troutbeck, 29, suc-
cumbed to his injuries the next morning at Hamilton Military
Hospital. Incredibly, two other mechanics aboard, AC 1 Lloyd
McLean and AC 1 George Sibley, walked away unscathed.
Sibley, who was on his first airplane ride, had volunteered
for the flight which was to take them over the Niagara Peninsula

F/S George Troutbeck died following the crash of Bolingbroke 9964. PATRICIA TROUTBECK

and his hometown of St. Catharines. Though originally seated in the cockpit, he had been ordered back to the cramped, mid-upper turret by the higher ranking Cpl. Doan—a move, it turned out, that saved Sibley's life. McLean, too, had been lucky. Scrambling from the wreckage, they found the pilot unconscious on the ground and Doan dead in the seat Sibley had been forced to give up just minutes before. Dazed, Sibley started towards the airport where he met the fire truck and ambulances bound for the scene of the accident, smashing through fences on the way. He wandered back to the crash, where somebody recognized him and ordered him to lie on a stretcher. He was

Above and facing page: Two views of the Bolingbroke crash that claimed the lives of George Troutbeck and Bill Doan. Two others somehow managed to escape. JOHN C. MILNER

F/S George Troutbeck is laid to rest with full military honours in the airman's plot, Knox Presbyterian Church cemetery, Jarvis. ART RIMMER

rushed to the station hospital where not so much as a single bruise could be found anywhere! He flew again, two weeks later.

Throughout the war, organizations within Haldimand and Norfolk were encouraged to include servicemen and women in their social events whenever possible. Occasionally, however, this led to problems. A highly publicized incident involving beer at a Simcoe high school dance on January 21, for instance, created hard feelings between irate parents of teenage girls and members of the RCAF. It was one of several such cases involving liquor at school dances and a precedent had to be set. Taken into custody and charged with having beer other than in his private residence was a

Sgt. Edmon Ryerse, air bomber, came from nearby Port Dover. EDMON RYERSE

Sgt. Cecil Bradburn, wireless operator/ air gunner, was from Janetville, Ont. JOAN STEPHENS

young New Zealand airman who confessed to having no knowledge of the Ontario Liquor Act. A lenient magistrate let him off with a minimum $10 fine accompanied by a stern warning, one that echoed all the way back to the base.

Due to poor weather conditions in January and February, flying was greatly curtailed, resulting in most of the courses being extended. Just over 200 students graduated from the station during those eight weeks, far fewer than usual. On February 9, the station's high level bombing flight, consisting of six Bolingbrokes under the command of F/L M.T. McKelvey, left on two months temporary duty to train high level air bombers at the No. 7 Bombing and Gunnery School in Paulson, Manitoba. On the 17th, the Canadian Red Cross mobile blood donor clinic visited the station, where 100 officers, airmen and women donated badly needed blood to the war effort. This was the first time that an RCAF unit in Canada had given blood on a station basis.

Situated 70 miles due west of Buffalo and in a direct line with Detroit, the station frequently received unscheduled visits from American crews flying over Canadian soil. During any given week it was not uncommon for US military aircraft to buzz the field or to land occasionally to make repairs. A wide array of frontline fighters and bombers, including P-51 Mustangs, Lockheed Hudsons, Venturas, B-25 Mitchells, Aircobras, Dakotas and C-46 Commandos all made appearances above the field, much to the delight of base personnel.

Sgt. (later Flight Lieutenant) Ted Potter, staff pilot, was from Beamsville. TED POTTER

It was during one such excursion on March 26 that a P-39 Aircobra crashed 11 miles southwest of the station near Port Ryerse. The pilot, Lt. John Anderson of the US Army Air Corps, bailed out too late and was killed when he struck the ground. Due to regulations governing those serving with US forces, his body could not be returned immediately to the States. For this reason he was buried with full military honours four days later in the Knox Presbyterian Church cemetery at Jarvis. Among those in attendance were six fellow American officers from Romulus, Michigan.

The spiritual welfare of those serving at air training schools throughout Canada during the war was made a priority by the RCAF. While married airmen and their families could, and often did, attend local churches of their choice, suitable arrangements for worship had to be made for the hundreds of others living in quarters on the base. Like most air force stations, Jarvis offered church services each Sunday morning and evening for those not flying or assigned to other duties.

Providing spiritual leadership at the school were two chaplains, one Protestant, the other Roman Catholic. In addition to conducting regular services and daily Mass, the chaplains assumed a variety of duties which included ongoing meetings with troubled or homesick personnel, presiding at wings parades, visiting those in hospital, performing the occasional wedding and addressing each new class upon arrival. It was no small task facing the chaplains with the total number of men and women on the station now surpassing 1700. Tak-

F/O Tom Kernaghan, air observer (instructor) hailed from Hamilton.
TOM KERNAGHAN

ing on this challenge in the spring of 1943 were F/L Richard Davidson, Protestant and F/L C.A. Doyle, Roman Catholic. Both padres, in their respective reports to the commanding officer, indicated a sharp decline in church attendance while at the same time reporting an increase in the number of airmen and women seeking counselling. Problems were often of a domestic or financial nature, made worse by growing frustration and uncertainty about the future. Three years of war, it seemed, was taking its toll.

In May the station's medical officer and marine section assisted in rescuing the crew and passenger of a Hagersville Anson which was forced to ditch in Lake Erie near Long Point. Notified that a plane was down, the bombing and gunnery school sent two crash boats to the scene. The rescuers arrived to find all four occupants and a Port Rowan man on the beach, wet and exhausted but otherwise okay. Just an hour before, 41-year-old Bill Mussell, manager of the Long Point Company, had witnessed the crash and set out immediately with the company launch and skiff to the partly submerged plane, located a mile from shore. After travelling as far as he could by launch, he took to the lighter skiff, paddling through the marsh until he reached solid ground. He then dragged the 200-pound boat 300 feet across land before entering the water again. At the scene he found three airmen and an airwoman sitting on the fuselage and quickly assisted them into the 14-foot skiff, which was already riding well down in the water. Fortunately the lake was calm, otherwise the small boat would have swamped immediately. Had it not been for Mussell's quick action and endurance, all four members of the crew might have been lost as the plane sank shortly after they were taken off. For his resourcefulness and determination, Mussell was awarded the Order of the British Empire. It would not be the last time civilians would come to the rescue of downed flyers in the lake.

Another harrowing experience—this one involving two Dunnville aircraft—occurred two weeks later. The incredible adventure of Sgt. Bill Hill and his instructor P/O Bill Bouton was later published in *Mentioned in Dispatches*.

RCAF marine crash tender M301, seen here departing Port Stanley. These high speed launches were kept at the range and could reach a downed plane in minutes. CWH/ A. JUDD KENNEDY

Sgt. Bill Hill following the famous parachute-cutting incident at Dunnville in June 1943.
NO. 6 SFTS ASSOC.

This pupil (No. 6 SFTS) should now by rights be in heaven, or wherever good little students go after a tour of duty on earth.

But he isn't!

And the only reason he's still with us is that he happened to lean forward slightly in his cockpit when he turned his head around to the front.

His instructor was giving him some dual on formation flying. They were formating with another Harvard carrying an instructor and pupil.

This other Harvard pulled alongside our student's plane in echelon port. Then, without any signals, whipped over to form echelon starboard.

Then "it seemed to hit an air pocket" and dropped toward our pupil's plane.

Said our pupil, "I looked around to the left with the intention of clearing my area, then signalled a turn to the left. When I turned my head back to the front, I leaned forward slightly and at that instant I FELT THE PROP OF THE AIRCRAFT HIT MY STRAPS BEHIND ME!"[1]

Hill at the time was flying with the front canopy slid open, which allowed the propeller from the aircraft passing overhead to enter the cockpit cleanly between Hill's shoulders and the seat! When Hill pushed the stick forward, F/S Jim Buchanan, instructor in No. 2 Harvard, pulled his back resulting in everything aft being mangled by the propeller. Luckily,

A bird's-eye view of the AMBT loft and student's position (l). The moving scenery is visible two storeys below. An air observer (r) peers down at the moving landscape of the Air Ministry Bombing Teacher NAC PA166778/DND PL 3575

Hill's instructor, who was seated in the rear cockpit, saw the propeller coming and ducked. Unable to see through the shattered coupe top, he screamed at Hill to take control of the aircraft, which was now in a spin. Even with the tip of the tail fin missing, Hill was able to level out and land back at the field without the use of flaps, touching down at 124 mph. At the station hospital, Hill was found to have three minor cuts on his back, requiring only a few stitches. Buchanan, who likewise managed to return safely despite a damaged prop, flew again that night.

Aircrew were not the only ones at risk. On June 4 two armourers, LAC C.A. Armstrong and AC1 J.W. Mack, were placing 11½ pound practice bombs in the bomb bay of an Anson when one exploded. The men were rushed by ambulance to the station hospital, where fragments were removed from Armstrong's left leg and Mack's abdomen and right leg. In an earlier accident, Harry Bullock, a civilian employed in the airmen's mess, was struck and killed late at night by a station vehicle as he walked along the Nanticoke side road. Still later, 13-year-old Howard Franklin was knocked off his feet when a stray bomb blasted a six-foot-deep hole in the sand 25 feet from where the boy was standing in Avalon Park, west of

In-flight instruction in an Anson. Bombing instructors from the plotting office often accompanied students over the target. NAC C74510

Port Dover. Such were the hazards of war, even in the relative security of the home front.

During the first week of June a new 24-hour RCAF training record was set by the station, with 127 hours flown by bombing flight, 197 exercises completed and a total of 1,241 bombs dropped. The bombing of stationary targets was normally done from altitudes of anywhere between 1,500 feet (low level) and 5,000 feet, with high level exercises carried out at 6,000. To climb much higher, the lumbering old Ansons would have taken forever—a luxury that a fast-paced training schedule could ill afford. To help get around this, the bomb aimer's actual errors at 4,000 to 6,000 feet were often recalculated to 10,000 feet, giving ground plotters a better indication of how students would have scored had they actually bombed from the higher altitude.

LAC Tom Lawrence, RAF, was an eager 20-year-old in training that fall and recounted his experiences and the method used:

The bombing pattern at Jarvis was in the form of a four leaf clover. During each run over the target the bomb aimer, lying

Armourers prepare to install practice bombs onto the wing racks of a Fairey Battle. NAC PA190917

prone in the nose, directed the pilot so that the target came down the drift wires of the bombsight. Corrective wording was right (one word) which signified for the pilot to turn 5 degrees right, after which the words steady, steady were repeated as the target came into view. Corrections left were left, left (two words) in case the intercom wasn't perfect. On each pass over the target the pilot made a 90 degree turn to starboard which created the four leaf pattern. The clover was repeated a second time for the always eight bombs.[2]

Once a student had factored in the speed and direction of the wind and worked out a heading, bombs were dropped one at a time over the target as the aircraft approached from different directions. Sometimes only one pupil was aboard for an exercise while other times two or more took part. When multiple students flew together each took his turn making calculations and releasing his allotment of bombs.

LAC Edmon Ryerse of Port Dover likewise trained as an air bomber at Jarvis in the summer of 1943. His logbook indicates

The hospital staff in 1943. The number of men and women employed at the 34-bed hospital averaged between 18 and 22. JARVIS PUBLIC LIBRARY

he flew a total of 36 hours and 15 minutes both day and night between July 20 and September 14. During that time he dropped a total of 86 practice bombs and fired some 1600 rounds of ammunition from the turret of a Bolingbroke. He received several more hours instruction at Ground Instruction School, the 200-yard machine gun range at Hoover's Point and on the Air Ministry Bombing Teacher (AMBT).

The AMBT was an ingenious device designed by a British inventor and housed in a special two-storey building at the school. Intended to simulate actual bombing conditions, it consisted of a loft containing a fixed bombsight and related instruments, film projector and large, horizontal screen depicting enemy territory. Lying on his stomach, the student was instructed to make all necessary calculations, then seek his target on the imaginary landscape projected onto the circular screen directly below. As the scenery moved past with its farms, towns and factories, he could watch the target as it came into view between the drift wires of the bombsight before pressing the release button. For several seconds the scenery on the screen

Two nursing sisters pose beside one of the school's two Ford ambulances. JARVIS PUBLIC LIBRARY

would continue to move, then stop where the "bomb" had hit. The instructor could then chart the pupil's score and assess his progress. As was the case in real bombing, close grouping of the bombs was what the instructors were really looking for, not whether they actually hit the target. In fact, at any of the schools, direct hits on any of the targets—real or otherwise—were few and far between.

The plotting section of a bombing and gunnery school was key to the success of the station's bombing activities. At Jarvis, the high standard of efficiency achieved was due in no small part to the close cooperation that existed between the staff of the plotting office, bombing range, chief instructor, Ground Instruction School, bombing leader and the AMBT. Members of the women's division, who formed a large nucleus of the staff in 1943, received constant refresher courses in the theory of bombing, bomb plotting and bombing analysis in order to better assess plotting charts and scores. The plotting office was also a place of instruction. To analyze more clearly individual strengths and weaknesses, a number of bombing instructors were assigned to the section, ready to accompany trainees over

Bolingbroke 10058 following a category B crash east of Cheapside on August 5, 1943. Someone placed a tarp over the turret, presumably for security reasons.

the target and to assist them as necessary, a factor which considerably helped to improve overall bombing scores.

As the school approached its third year of operation, it continued to hold regular wings parades every two weeks, often with dignitaries present. Saturday, June 12 was just such a day when, before the whole assembled station, members of Course 54 Wireless Air Gunners received their wings from Air Marshal William Avery "Billy" Bishop, VC, DSO, MC, DFC. "It is daily apparent what an important part the air is taking in this war," the air marshal began.

Without complete supremacy we cannot hope to win. We have it now and I assure you that we do not

Bolingbroke 10058 was shipped to Trenton by a mobile repair party from No. 6 Repair Depot.

A graduating pilot receives a congratulatory handshake from Air Marshal Billy Bishop. Cloth propeller badge on sleeve denotes rank of leading aircraftman, white flash in cap indicates aircrew trainee. DND PL 5034

intend to lose it. The job ahead of you is not an easy one. It is going to be hard and hazardous but on you depends the winning of the war which is not yet one. You will meet a fierce and fighting foe ... but you will meet him with confidence that you are better trained than he is, that you are better equipped, that you know your job better and that you are a better man than he is. You will win, never fear. Your job is just as important as that of the pilot or navigator. On you, as much as them, depends the safety of the rest of the crew, of the aircraft and the successful conclusion of your mission.[3]

At his own insistence, A/M Bishop frequently officiated at such ceremonies across Canada. Bishop,

Bolingbroke cockpit. Bomb aimer's compartment is clearly visible on the right. TED POTTER

Seven young men died when two Ansons collided over Oneida Township on June 15, 1943. F/O E.B. Norbury (top left), pilot of Anson 7566, was killed on impact, along with students LAC J.H.Kearney (top centre), LAC J.A. Smith (top right), and LAC H.P. Samuel (second row, left). LAC A.R. Hayes (second row centre), a student in Anson 7339, was killed outright; fellow passengers LAC F.E. Innes (second row right) and LAC J.A. Holgate (bottom right) were fatally injured. Miraculously, the pilot of 7339 survived. NAC 144230/C144227/C144235/C144233/C144224/C144226/C144225

a native of Owen Sound, Ontario, became director of recruiting for the RCAF in January 1940. He survived more than 170 air battles in the First World War and was officially credited with shooting down 72 German planes. Everywhere he went servicemen and women were thrilled to hear him speak and some were even lucky enough to have their wings presented to them by the legendary ace.

The wings parades, normally held in the presence of the entire station and often attended by relatives, were colourful and dignified affairs. As each graduating student's name was called he would proudly step forward, march to a table containing the coveted badges, then snap to attention before the air marshal. Putting each man at ease with his reassuring smile and chatting briefly with him, Bishop then pinned the student's wings on the left breast of his tunic. Following the usual congratulatory handshake and expressions of good luck from the air marshal, the student then stepped back, saluted and returned to the ranks.

While en route to the school's B.5 land target near Willow Grove on June 15, two Ansons collided at 2000 feet over the Township of Oneida. The collision occurred at 1850 hours, approximately 15 minutes after takeoff. At this particular time, P/O Ralph Herring in Anson 7339 was apparently flying straight and level and F/O Ed Norbury in aircraft 7566 was coming up in a climbing turn, with the intention of passing behind him. Through an error in judgment, his port wing and engine collided with 7339 just forward of the tail assembly. Trailing smoke, Norbury's aircraft plunged to the ground and exploded on impact, killing the 21-year-old pilot from Windsor, Ontario along with Canadian students John Kearney, James Smith and Hector Samuel. At the same time, Herring's plane, minus part of its tail, began a crazy spiral downward, hitting the ground a mile away. One student, LAC Arthur Hayes was killed outright while another, LAC Fred Innes, was carried to the living room of a nearby farmhouse where he died 20 minutes later. P/O Herring and a third student, LAC John Holgate, were rushed unconscious and in critical condition to the hospital at No. 16 SFTS, Hagersville where Holgate died of his injuries on June 18. It wasn't until early July that Herring had recovered sufficiently from a broken back and concussion to speak for the first time about what had happened. On July 4 he issued the following statement:

> On the evening of June 15, 1943 I was detailed to do Bombing Exercise B3 over B.5 target, at 6000 feet. I very vaguely remember taking off and can remember nothing further regarding the flight except a vague recollection of falling. The plane seemed to be going straight down and all the crew seemed to have been thrown forward to the cockpit. Next thing I remember was lying in the hospital.[4]

Against all odds, Herring survived to fly again.

Having established a new bombing record that June, it was not surprising that bombing, gunnery and drogue flights logged the most hours ever flown at the school. Staff pilots were kept busy as they chalked up a total of 4,620 hours that month—the equivalent of having a plane in the air continuously for more than half a year! Flight schedules were demanding and at times gruelling, particularly when pilots were required to make five or six trips in one day. In addition to the routine bombing and gunnery runs, pilots were assigned a variety of duties including wind-finding exercises, test flights, weather checks, navigation exercises and the occasional cross-country trip to places like London, Trenton or Camp Borden. Flights lasted anywhere from 20 minutes to three-and-a-half hours.

Needless to say, traffic both on the ground and in the air was heavy. One after another, Ansons and Bolingbrokes taxied to active runways, awaiting the signal for take-off while incoming planes rolled off the landing strip and onto the flight line, staying just long enough to discharge one crew and take on a new one. At the same time, armourers swung into action, replacing bombs in the bomb racks while gunnery students

1943 - In Full Swing

American visitors always attracted attention. Left, a P-39 Aircobra of the U.S. Army Air Corps; right, a Lockheed Hudson.

reloaded machine guns and fuel trucks replenished gas tanks as necessary.

Sgt. Ted Potter of Beamsville, Ontario was typical of the 20 or so pilots attached to bombing flight. At 25 years of age he had been posted to Jarvis twice, first in 1941 as an aircraftman 2nd class and again, upon request, in July 1942 as a new sergeant pilot. He well remembers his daily sojourns to the flight office in Hangar C, perusing the duty board, signing out aircraft and chatting briefly with the students who accompanied him over the target. Teamwork between the pilot and students was essential, despite the fact they seldom flew together more than once or twice. It was no secret that a good pilot contributed to better bombing scores and Potter, whose hours per month were nearly always top in bombing flight, was considered among the best.

Unlike the majority of staff pilots, who found the job dull and often times monotonous, Potter thoroughly enjoyed what he did. Except once. On a cold, bitter night in January 1943 he and one other pilot were sent up in an Anson to complete a quick 20-minute weather check. Although it was clear upon takeoff around 8 o'clock, the weather quickly changed, catching the two men alone in near-zero visibility in heavy snow. After radioing the tower that they were returning to base, they found themselves hopelessly lost in the worsening storm. Descending to a few hundred feet in hopes of pinpointing their location, they successfully identified the town of Waterford by its lights, then proceeded in a southeasterly direction on a path they knew would eventually lead them over Port Dover. After locating the familiar resort town a few minutes later, they plotted a course back to base and landed on snow-covered runways nearly an hour after take-off, tired and shaken.

Fortunately, life at the school was not all work. There were supervised swimming parades to the lake for aircrew trainees each night during the summer, along with a whole list of recreational activities at

Drogue operator Ian Moon of Hamilton. The tedious job of unfurling drogues and operating a winch–while seated backwards–was an unpopular assignment and never received the recognition it deserved. IAN MOON

the station. A fully equipped sports field, prepared by works and buildings and located west of the maintenance hangars, provided first-class playing conditions for most track and field events and included an oval quarter-mile cinder track. Team sports, too, played a key role in keeping men and women physically fit and helped occupy leisure time. Major sports included softball, basketball, hockey, soccer, volleyball and football, while tennis and badminton also proved immensely popular. All available courts were in use most summer evenings beside the drill hall.

To accommodate such an ambitious sports program an inter-unit league, similar to intramural sports in high school, was formed. This league, consisting of 14 teams for each sport, represented all six hangars, six bomb aimer courses, wireless air gunners, administration and ground instruction school staff. Players who exhibited exceptional athletic ability were quickly promoted to the school's more prestigious inter-station league, whose teams competed regularly against neighbouring airfields at Hagersville, Dunnville, Brantford and Mount Hope, as well

F/S Harry Head instructing gunnery student at No. 1 Bombing and Gunnery School, 1943. NAC C74463

as the No. 25 Canadian Army (Basic) Training Centre in Simcoe.

Rivalry between stations was keen. On August 17, the women's division softball team soundly defeated Hagersville 7-3 for the right to travel to the No. 1 Training Command play-offs in Toronto later that month. After eliminating Camp Borden on the 28th, they lost 8-5 to RCAF Aylmer in the final—their second loss of the season. On several occasions Jarvis teams earned their way to the command finals and more than once came away with the title.

On September 1 the station was host to yet another high-profile visitor when W/C Guy Gibson, VC, DSO and Bar, DFC and Bar, arrived at Jarvis, following an official itinerary arranged by air force headquarters. Gibson had been awarded the Victoria Cross, the Commonwealth's top military award for valour. He also received the DSO for a daylight raid

W/C Guy Gibson, VC, DSO and Bar, DFC and Bar, with G/C A.D. Bell-Irving, September 1, 1943. Gibson led the famous Dam Buster raid of May 1943. NAC c74462

on Milan and won the Bar for another daylight raid on Stuttgart. He won the DFC for attacking invasion barges off the coast of France and Belgium and the Bar was awarded for shooting down four German aircraft over England in two nights. At just 25 he had already participated in 175 operational sorties including 76 bombing raids over Germany.

Gibson is no doubt best remembered for his role in leading the famous Dam Busters raid on three strategic dams in Germany just three-and-a-half months earlier, using "bouncing" bombs which skipped across the water before detonating against the walls of the dams. For his part in the operation, Gibson received the Victoria Cross. Nine other decorations went to Canadians involved in the operation. It seemed particularly fitting that the station was included on Gibson's four-week, 12,000-mile cross-Canada tour since some of the Lancaster crews hand-picked for the daring raid, including 30 Canadians, were themselves graduates of Jarvis and surrounding schools. Gibson's navigator for the raid, F/O Terry Taerum from Milo,

Refueling a Lysander. The station had storage capabilities for 20,000 gallons of gasoline.
CHAA

Alberta and air gunner WO II Albert Garshowitz of Hamilton, both received their training at Jarvis. Garshowitz was one of 13 Canadians killed in the raid.

To suitably welcome such a distinguished officer, the station pipe band led a parade along the main roadways of the school, where virtually everyone on the base had turned out to greet him. Following a luncheon in the officer's mess, the wing commander lectured the staff and trainees on operations overseas:

> The day is gone when you work as an individual and now all members of a crew in a bomber are part of a team. In bombing raids, the crew now comprises an offensive and defensive team, working in one complete unit. Each member of the crew is just as important as the pilot and they all depend on one another during the entire raid. It is almost a certainty now that if a pilot receives a decoration so will all members of the crew of his aircraft as they work as one complete unit.[5]

Following his address, G/C Bell-Irving presented W/C Gibson with a sweater bearing the crest of No. 1 Bombing and Gunnery School.

Graduates of Moving Target Course 81, December 13, 1943.

In recognition of the significant role the school played in training Polish airmen, G/C Bell-Irving was presented with a Polish air force pilot's badge at a special ceremony following the weekly station parade on September 18. The presentation was made by G/C S. Sznuk, official representative of the Polish air force in Canada.

"It just so happens that this training establishment is one at which many of our Polish airmen have been trained for the past year or so," began G/C Sznuk in his address to the personnel of the station.

> Out of this school, under your command Group Captain Bell-Irving, our men not only graduate in the art of air gunnery and bombing an enemy target, but they also carry with them the precious and unforgettable expression of the perfect hospitality of this land and its people.[6]

G/C Sznuk concluded by saying that Polish airmen deemed it an honour to fight beside Canadians and with those remarks, pinned the Polish wings on G/C Bell-Irving.

Beginning October 18, the fifth Victory Loan drive got underway with the school's share of the quota being set at $80,000. In an effort to increase this amount even more, Jarvis was entered into competition with the flying schools at Dunnville and Hagersville as well as against the county of

Haldimand. The main objective this time around was the German submarine base at Kiel. The purchase of bonds by county residents was to provide funds for Mosquito bombers, Dunnville and Hagersville would supply the trained crews and Jarvis the bombs. To inform personnel of the progress, a large chart was erected outside the drill hall, where sales were posted daily. To kick off the campaign, the women's division pipe band led a Victory Loan parade in Cayuga, the county's central town.

Late fall brought with it a number of significant changes and innovations. G/C Bell-Irving, who had served as commanding officer since August, 1942, was posted to Central Flying School at Trenton to act as CO there. He was replaced by W/C William Peace, DFC of Hamilton, who assumed temporary command at Jarvis. Other changes included the expansion of the marine section at Port Dover, extension of No. 3 runway to 4200 feet, completion of a new women's hairdressing salon in the recreation centre and the introduction of the "Jarvis" bomb. Owing to the shortage of government-supplied practice bombs, this new bomb was designed and developed by the armament staff and manufactured locally. Its use for daylight bombing allowed the limited number of issue bombs available to be used for night operations.

On November 19 Lysander 2315 was forced down at Dunnville when the propeller and reduction gear flew off in mid-air. The drogue operator parachuted to safety while the pilot glided to the Dunnville aerodrome, where he made a perfect landing. It was a much different story three weeks later when another Lysander developed engine trouble, crashed into a creek bank and exploded four miles southeast of the station. Both crew members bailed out just in time, the pilot receiving a broken ankle.

One of the largest wings presentations ever to take place at Jarvis was held on December 11 when over 100 airmen graduated as air bombers, air navigators B and wireless air gunners. On hand to capture the ceremony on film was the crew of Associated Screen News, who were preparing a full-length documentary on the RCAF Women's Division. Countries represented that day included Canada, Great Britain, Australia, New Zealand and Norway. By this time the school had realized its full potential, with the number of officers, trainees, airmen and airwomen peaking at 1,857, including 147 civilians.

With so many young men and women brought together in one place it was inevitable that friendships would develop both on and off the base. Not surprisingly, more than one member of the women's division accepted marriage proposals from airmen they came to know while handing out parachutes, driving one of the fuel trucks or serving in the mess. In fact, in its final report in 1946, the BCATP supervisory board confirmed that nearly 3,800 Commonwealth airmen sent to Canada during the war had indeed married Canadian girls.

Still others, with similar ambitions, never got the chance. One such couple was Enid Williams of Simcoe and P/O Doug Guest, a 21-year-old Australian pilot who arrived at Jarvis in March 1943. They met, as many had, on a prearranged date which Enid fondly recalled.

On April 8, 1943 I was introduced to Doug Guest by another pilot I had gone out with a few times from the Jarvis Bombing and Gunnery School. I phoned a girlfriend to make a foursome and we went bowling in Delhi. Once or twice after that the four of us went to the movies then Doug started phoning me on his own and came in every free night he had and on 48 hour leaves too. He would come to Simcoe from the airport on a special bus for the airmen at 60 cents a night. The bus driver got to know Doug very well and if he was late getting downtown to the bus stop at the end of the evening, the driver would start up Colborne Street to meet him. I lived a mile from downtown so by the end of some evenings he would have walked at least four miles—one mile to call for me, one mile back downtown to the theatre or a restaurant, back to my house then back downtown to catch the bus. We had lots of time to get to know each other just walking and talking.[7]

The two became best friends. They played golf, went on picnics or rode bicycles to Port Ryerse, Waterford or nearby Lynn Valley. Sometimes they attended dances on the station or at the Simcoe high school auditorium on Saturday nights.

One Sunday, Doug invited me to dinner at the Officers Mess before taking me up in an Anson on a weather check. I had to sign a

Doug Guest and Enid Williams on a date in Simcoe. ENID BLUME

waiver but it took a lot of running around before somebody found one. Then, when they put a parachute on me, I refused to do it up properly as I was wearing a dress. Very embarrassing! Doug always said it was the best entry in his log book and that I was his favourite co-pilot.[8]

New Year's Eve 1943 found Doug and Enid alone in her sister's house, staring at the fireplace when Doug suddenly got the idea of placing a lock of Enid's hair behind the pilot wings on his tunic. They carefully removed the wings, padded them with hair and were struggling to sew them back in place with her sister's machine when she came home and offered to help. She was surprised, to say the least, but seemed to understand what they were doing and why and helped finish sewing them on.

Like most of the boys, Doug was keen to get overseas and finally got his wish three months later. As he had done in Canada, Doug wrote Enid constantly, usually every other day. When he wasn't writing he learned to fly Spitfires and later Typhoons and it was in the latter that he struck a high-tension cable while on a strafing run in Europe and was killed on February 13, 1945.

P/O Doug Guest, Royal Australian Air Force. ENID BLUME

Upon being told of his death by her mother, Enid hastily penned the following words, which remain as poignant today as they were back then:

> The love that comes to many not at all
> Came to us while we were young.
> The love that comes but once, was ours—
> but not for long.[9]

Chapter Five
1944 - Leveling Off

Early in the new year, foolhardy flying at air schools throughout the country became an area of concern once again. On January 11, G/C F.S. Wilkins, RCAF chief inspector of accidents, appealed nationally to the general public to regard stunting and low flying pilots as "dangerous drivers" and implored citizens to, "Report them—don't wave to them." Wilkins, who had been investigating air crashes since the start of the war, stated that too often accidents were the result of show-off tactics by inexperienced pilots or bored instructors. Strict rules governing the practice of aerobatics stipulated they were always to be performed at high altitude and never over cities, towns or villages.

"Civilians who watch, heart in throat, can easily be the means of saving a life by noting and reporting the number of the plane,"[1] the group captain emphasized. If the number could be read easily from the ground, authorities explained, then the plane was too low. While all schools were guilty of violations, some were worse than others. Pilots attempting a roll at low altitude, buzzing a girlfriend's house, flying up Main Street or zooming under bridges and high tension wires were all courting disaster. The numerous incidents of low flying, carelessness,

RCAF photographers from bombing and gunnery schools right across Canada attended a familiarization course at Jarvis, January 31, 1944. RUTH STUART

disobedience and pilot error ultimately contributed to the majority of mishaps and many of the 856 trainee deaths which occurred in the nearly five-and-a-half years the training scheme operated in Canada.

In February, staff and students at the school welcomed G/C W.J. "Packy" McFarlane as its latest commanding officer, replacing Acting CO W/C Bill Peace. Prior to arriving at Jarvis, the popular 40-year-old McFarlane had served as commanding officer of the Northwest Staging Route, headquartered at Edmonton. A mining engineer and Vancouver city policeman before the war, he had also commanded RCAF station Goose Bay, Labrador and opened the bombing and gunnery school at Picton, Ontario.

On March 29, P/O W.B. Townley, a respected instructor on staff at the school, was presented with the Distinguished Flying Medal at the weekly commanding officer's parade. Townley, who had served with distinction as an air gunner with No.97 Squadron, RAF, had recently been promoted and returned to Canada. The citation to his medal, read by G/C McFarlane, stated:

Anson 7043 following its miraculous landing minus eight feet of wing!

Sergeant Townley, as air gunner, has completed a large number of operational sorties. His targets have included attacks on some of the most heavily defended objectives in western Germany and Italy. He has always executed his tasks with great coolness, courage and efficiency.[2]

Coolness and courage took another form closer to home three weeks later. Just before midnight on April 21, Sgt. John McRae of Vancouver and two students were flying on a routine bombing exercise over Lake Erie when suddenly an Anson on a weather check from Hagersville appeared out of the dark. Both pilots immediately took evasive action and as a result the crew of Sgt. McRae's plane felt only a slight jar. On making an inspection with his flashlight a few seconds later, however, the pilot discovered to his horror that he was flying with eight feet of his starboard wing missing! He headed his aircraft toward shore and once over land gave LAC's Cahill and Armes the order to bail out. Both students landed safely approximately a mile apart and were picked up by members of the motor transport section. The pilot, however, was unable to follow, as every time he let go of the controls the Anson threatened to go into a spin. "So I hung on," said McRae, "and managed to make it back to the airfield."[3]

Displaying keen airmanship, he used what little control was left and made a successful crash-landing. When fellow pilots and technical officers examined the aircraft in daylight, they said his feat was nothing short of a miracle. As a result of his actions, McRae, 25, was recommended for the Air Force Medal.

Needless to say, both Cahill and Armes jumped at the opportunity to abandon the crippled plane. Like all aircrew, they were grateful for that small bundle of silk they carried with them in the event they needed it. Pilots, who went aloft several times a day, wore a standard seat-type parachute which doubled as a cushion in the hollowed-out seats of most training aircraft. Students,

Don't wanna forget this! During training flights it was strictly forbidden to go aloft without a parachute. Shown here is the chest-pack type used by students. DND PL2212

meanwhile, carried aboard the more compact chest-type pack or quick-connector type, which was stowed in the fuselage and kept close by. These chutes could be clipped easily onto an airman's harness in an emergency and released instantly when landing in water.

Hangar F. Raised part of lean-to at left houses two-storey loft where parachutes were hung to dry. MICK HENNING

Leading Airwoman Irene Osborne of Dunnville spent three years in the parachute section at No.1. Like a number of girls, she met and married Joseph Goldspink, one of the students she came to know while assigning parachutes for training flights. She remembered Jarvis as a well-run station with good

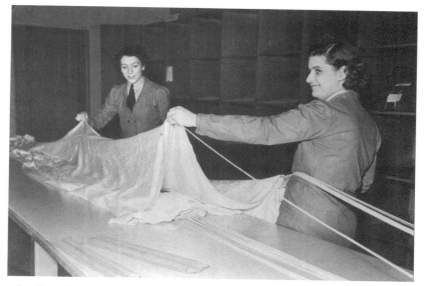

Leading Airwomen Doris Hennessy (l) and Irene Goldspink carefully fold the silk canopy of a parachute. IRENE MILLER

orchestras, summer wiener roasts on the beach and dances at least once a month. The latest movies from Hollywood, shown in the recreation hall, cost 25 cents.

"When we first arrived we were called fabric workers," she recalled, "We mended the aircraft then put three coats of dope over the repairs. We also packed parachutes. Later, it became two trades — fabric worker and parachute rigger."

Packing a parachute took two people about three-quarters of an hour. Each panel of the silk canopy was carefully inspected, as were each of the 24 shroud lines. Parachutes were unpacked and inspected once a month whether they had been used or not and hung to dry for 48 hours in a chute well — a two-storey loft located in F hangar. This process helped eliminate moisture and remove creases. A log was kept for each parachute and the girls had to sign in three places that they had inspected it.

"I remember one incident in particular," related Irene with a chuckle years later.

The Cornell trainer purchased for and donated to the RCAF by Simcoe elementary and high school students. NORM WILLIAMS

The sergeant of our section was in the mess one night when one of the fellows approached him and said in no uncertain terms that he hoped he would never have to jump out of a plane with a parachute packed by a woman. Much to his surprise, that same pilot had to jump the very next day and was grateful to receive only a broken ankle![4]

In May the station was awarded the highly sought-after minister's efficiency pennant for the first quarter of 1944. This pennant, awarded regularly to the bombing and gunnery school deemed most efficient in Canada, was proudly hoisted beneath the RCAF ensign on the flagpole at the east end of the parade square. Jarvis won it again for the second quarter of 1944.

On May 6 a new 40-foot twin-engine cruiser was added to the growing fleet of boats attached to the marine section and that night a dance was

Learning how a parachute behaves.
JOHN C. MILNER

Polish personnel attached to No. 1 Bombing and Gunnery School, May 8, 1944.

held in honour of the second anniversary of the women's arrival at Jarvis. One month later, on June 6, the long awaited Allied invasion of Normandy got underway with all personnel glued to every available radio on the base. At a special church service held in the afternoon prayers were offered for all friendly forces taking part in D-Day operations.

The next day, children of the various schools in Simcoe donated a brand new Cornell trainer to the RCAF. Accepting the aircraft on behalf of the air force was No.1 Bombing and Gunnery School. Approximately 800 students and 200 other guests were present at a special ceremony held on the station along with the Simcoe High School army cadets, Simcoe air cadets and members of the high school band.

Arriving on schedule at 3:30 that afternoon, the factory-fresh Cornell circled low above the field with F/L Bill Whitside, himself a former Simcoe student, at the controls. Whitside taxied to the platform set up for the occasion, where he was greeted by G/C McFarlane, former Ontario premier Mitchell Hepburn, Simcoe mayor Bruce Whitside—Bill's father, members of the school board and representatives of all six schools participating in the campaign. At the conclusion of the ceremony, G/C McFarlane expressed appreciation on behalf of the air force, after which he presented a plaque to the Simcoe students—the first in Canada to reach their objective of buying a plane for the RCAF.

The purchase of the aircraft, affectionately named *Town of Simcoe*,

Air Cadets work an Aldis lamp atop the control tower at Dunnville. TREVOR MELDRUM

was made possible through students' purchases of war savings stamps and certificates amounting to $11,600. During the war, savings stamps could be purchased for 25 cents each and were available at most banks, post offices and other sales agencies. For every 16 stamps collected, placed in a special folder and mailed postage-free to Ottawa, one five-dollar war savings certificate was registered in the purchaser's name and mailed to him or her. Children were among the most enthusiastic supporters of the war effort, purchasing stamps out of their allowance or with money earned from part-time jobs.

Teenage boys between the ages of 15 and 18 contributed to the war in other ways, too. Similar to the cadet corps or reserve units of the army and navy, the Air Cadet League of Canada offered young men an opportunity to see up close the work being carried out by the RCAF. In cooperation with the cadet league, flying schools right across the country offered summer camps for young cadets considering a future career in the air force. For the most part the boys were treated like those in the regular service, fed in the airmen's mess and quartered in tents or permanent barracks as room and space allowed. The main purpose of the week-long camps was to familiarize youngsters with most aspects of RCAF life, including drill, airmanship, navigation, wireless, study of aero engines, armament, signals, aircraft recognition and theory of flight.

1944 - Leveling Off

By the summer of 1944 arrangements were finally completed to give familiarization flights in RCAF aircraft to senior air cadets who qualified. As a rule of thumb, cadets received flights as a reward for faithful attendance at squadron parades and progress in training. Parents signed a waiver form for each cadet who qualified and was recommended for

LAC Jack Paton and wife-to-be Ila, both of Parkhill, Ontario, spend some leisure time on the beach in the summer of 1944. JACK PATON

flights by his commanding officer. Trips were of 30 minutes duration or less and were carried out within a radius of 20 miles from the station. Prior to this, air cadets had been prohibited from flying.

Time proved the cadet league to be a complete success. Every summer, upwards of 1,800 air cadets from as many as 27 squadrons enjoyed the hospitality extended to them by various RCAF flying schools throughout southern Ontario. A rigorous air cadet officer training program was offered at No.1 Bombing & Gunnery School for two weeks in July. By the spring of 1944, the Air Cadet League of Canada had established 371 active squadrons, comprised of more than 29,000 air cadets. Since the same time a year earlier, 1,906 former air cadets had joined the Royal Canadian Air Force, with almost as many enlisting in other branches of His Majesty's service.

In June 1944, after waiting what must have seemed an eternity, local commercial fishermen finally received their share of payments promised by the federal government as part of a compensation package for fishing grounds lost to the Lake Erie bombing range. An agreement, signed in 1941, provided $200 per net, per year, plus one payment of $200 per net at the end of the war, to assist fishermen in getting re-established and in buying new equipment to replace nets and twine which had deteriorated by lying idle. Major fisheries affected in the immediate area included those based out of Port Dover, Nanticoke and Port Maitland.

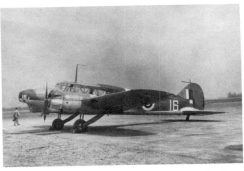

One of the Mk I Ansons shipped to Canada from the UK. Much to the consternation of aircrews the landing gear had to be hand-cranked 151 turns into the up position after every take-off.
MICK HENNING

That same summer, No. 1 Bombing & Gunnery School boasted another first with the installation of the synthetic trainer, a new, complex mechanism at the bombing teacher (AMBT). This device consisted of an enclosed, rectangular box large enough to accommodate a student air bomber. The interior featured a mock-up of the bomb-aimer's position found in

Part of an engine cowl and motor mounts from Anson 6086 which crashed near Port Dover on July 3, 1944. In 1999 the author, in company with Robert and Edmon Ryerse, unearthed dozens of pieces from the crash site. NORM WILLIAMS

operational bombers and contained a bombsight, switches and all other equipment necessary to guide and drop the deadly TNT carried below. Also incorporated into the trainer was a bomb-distributor or "Mickey Mouse" with which the student could make adjustments and drop his bombs at desired intervals. While the bomb aimer released his imaginary bomb load, moving pictures of enemy territory generated by the AMBT passed underneath, giving the student the effect of working from a moving aircraft. The realism of all this was intensified by the fact that while the practice was in progress, the machine gently rocked up, down and sideways, simulating an actual aircraft in flight and purportedly prompting an order for additional airsickness containers! To complete the effect, there was a position for the "pilot", who controlled the "aircraft", causing it to bank and turn according to directions given by the student air bomber, with whom he was connected by intercom. This piece of equipment was considered to be a great advance in the training of air bombers and the students training at Jarvis were the first in Canada to use the new device.

Crash scene of Lysander 2317 at Rainham Centre, July 31, 1944. It was believed the pilot had sufficient height to glide safely to the relief field at Kohler. CHAA

In the wee morning hours of July 3, residents of Port Dover were jolted from their sleep by the sound of a low-flying plane circling the town. Rushing outside, neighbours gathered in the streets and pointed to a plane trailing flames as it continued a slow 180-degree turn to the north. The aircraft, an Anson Mk I, had just cleared the target at Port Ryerse and was returning to base when the port engine burst into flames, forcing the

Australian F/S J.B. Watts was among those who died when an engine burst into flames in July 1944. RAAF

pilot to fly low over the area in search of a suitable place to land.

Hamilton resident Herbert Zimmerman was returning home on Highway 6 about 12:45 a.m. when the flash of a distress flare caught his attention. Pulling to the side of the road, he watched the plane as it banked steeply to the west, its left wing now totally engulfed in flames. The aircraft struck the ground at a sharp angle and at considerable speed in cartwheel fashion, with the port wing hitting first, followed by the port engine nacelle, nose section, starboard engine nacelle and starboard wing. When Port Dover chief of police J. Wally Montrose and others arrived at the scene minutes later, they found pieces of the plane burning fiercely in the field, along with the bodies of all three crewmen. The dead were F/S John Bradford Watts, 22, of Australia and students Gordon Best, 19 and Robert Waller, 21, both of England.

Daylight revealed the horrific extent of damage. Wreckage was strewn over a distance of 500 feet; the major portion of the fuselage had broken through a fence and burned in a lane. The motors, two 350 h.p. Armstrong Siddeley Cheetahs, had snapped from their mounts and rolled 900 feet into an adjoining field. Following a week-long investigation, RCAF officials speculated that a badly fitted propeller had vibrated to such an extent that a gas connection became loose and started a fire in the port engine nacelle.

An apparent engine failure four weeks later sparked even further controversy. While participating in a high-level gunnery exercise at 5000 feet, the motor of Lysander 2317 was heard by cottagers to sputter then cut twice. Quickly releasing the drogue and cable, LAC Stephen Garland attached his parachute and bailed out as instructed by the pilot, drifting north above the lake and landing on the roof of an abandoned barn just yards from the water's edge. The pilot meanwhile began a slow descent, circling the tiny hamlet of Rainham Centre, four miles east of Selkirk. With little or no power, the single-engine plane was seen to stop on a turn at 1500 feet and plunge straight down. Killed instantly was P/O Howard Green, the 19-year-old son of a United Church minister and youngest pilot on staff at the school.

P/O J.H. Green died when his plane crashed near Rainham Centre. NAC C144223

Testifying at the court of inquiry a few days later, one Jarvis pilot stated:

The body of P/O Howard Green lies at rest in the recreation hall. DR. BURDGE GREEN

I personally am not satisfied with the performance of any of the Lysanders I have flown at this Unit, and it is my opinion that this lack of good performance is due to longevity. I am of the opinion that these aircraft have been flown longer than they should have been and as long as they run up to the required revolutions we

Funeral procession for P/O Howard Green moves along the main roadway of the school.
DR. BURDGE GREEN

take them off. Their action with the target extended leaves plenty to be desired. I further state that it is not a matter of maintenance but of the gradual wearing out of the equipment.[5]

The investigating RCAF engineering officer could offer no explanation as to why the engine had suddenly cut out. Nor could anyone explain why the pilot had not landed without the slightest difficulty, considering the engine failed at 5,000 feet and the Lysander was designed to land in restricted places.

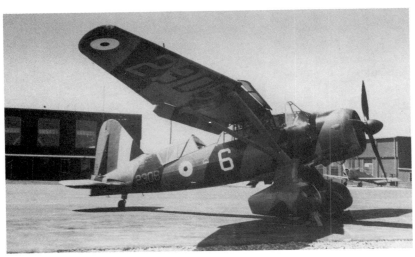

The Lysander III was powered by an 870 h.p. Bristol Mercury XX engine, had a top speed of 230 mph and a 50-foot wingspan.

The station glee club practiced twice weekly and sang regularly at concerts and church services in Simcoe, Jarvis and Port Dover. Seated front and centre is G/C W.J. McFarlane. THE FLY PAPER

It was a happier occasion on Saturday, August 16 when the school opened its doors to the general public for its first-ever Jamboree and Sports Day. The entire day's activities, which included sporting events, aerial displays and tours of the base, were arranged by permission of G/C McFarlane, with all proceeds going to the station sports fund. The purpose of the event was to acquaint the public with the training program which their Victory Bond purchases were helping to finance and an estimated 3,000 civilians from throughout the Niagara Peninsula took advantage of seeing an RCAF station on duty. Among the almost 3,000 service personnel also in attendance was Air Vice Marshal A.T.N. Cowley, air officer commanding No. 1 Training Command and G/C Gerald Nash, commanding officer of No. 6 SFTS, Dunnville.

The day began with a wide array of track and field events, followed at six o'clock by two softball games which pitted the Jarvis men's team against *HMCS Star* of Hamilton and the women's division against the girls of No. 4 Bombing and Gunnery School at Fingal. Immediately after the softball games, the crowd assembled on the fringe of the sports field to watch the inter-station soccer final in which Jarvis defeat Hamilton 2-0. Other attractions throughout the afternoon included guided tours, parachute jumps, a fly-past by three Mosquito bombers from Downsview and breathtaking aerobatics in a Harvard flown by P/O Norman Benbrough of Australia. Hot dog and Coca-Cola stands erected at various locations about the grounds provided refreshments all day long, with supper being provided for a minimal charge in the airmen's mess. As the outdoor events drew to a close, the crowds made their way to either the variety show at the recreation hall or to the drill hall for the sports dance, where music was provided by the popular dance orchestra, the Hagersville Ladies Starlettes. Prior to the dance, G/C McFarlane presented numerous trophies and awards to the winning contestants.

On August 25 another Haldimand air school found itself in the spotlight when 58 Canadian and Norwegian pilots graduated from No. 16 SFTS. On that day, members of Course 100, the longest in service flying history, received their wings under a new program called the combined training scheme. This was a totally new concept introduced at Hagersville, whereby students received instruction on both Harvards and Ansons, discarding the long-accepted RCAF method of carrying out advanced flying on one type of aircraft only. The reason for revision was simple. Four years experience overseas had proven that RCAF requirements often called for fighter pilots when only bomber pilots were available and vice versa. Moreover, some aircraft, like the Mosquito and Beaufighter, were twin-engine, which meant pilots needed to be conversant with the more powerful types.

"It definitely trains a better pilot", G/C Dave Harding, DFC, commanding officer at Hagersville said of the extended 28-week course. "He gets all his real flying, his aerobatics on a Harvard and all his school work, his navigation, on Ansons."[6]

Addressing an unusually large number of friends, relatives and newspaper reporters present on the parade square that afternoon, G/C Harding continued.

1944 - Leveling Off

They have done their job in a grand manner- probably as no other course has done. The class will make a name for itself wherever it goes. We switched them over to the combined training and they have been trained in both single and twin-engine aircraft, a training which will fit them for any job to come their way. With their wings goes the best of training given anytime, anywhere.[7]

The ceremony concluded with an impressive air display by Mosquito bombers from Downsview and Cornell trainers flown over from the Norwegian air force base at Little Norway, near Gravenhurst. Also brought in from Malton for the occasion was a four-engine Lancaster in which graduates were given a ride. Among those in attendance was Lieutenant Colonel Ole Reistad, air officer commanding the Royal Norwegian Air Force in Canada, who presented wings to Norwegian members of the class.

By September 1944 a surplus of aircrew saw the RCAF contribution to the British Commonwealth Air Training Plan reduced considerably. Owing to much lower than expected air casualty rates during the D-Day invasion, air force recruiting in Canada all but came to a halt and it was announced from Ottawa in November that the plan would be terminated in March of the following year. Everywhere, air training schools across the country began to disband and pilot recruits who found themselves on long waiting lists to begin their training were given the choice of re-mustering as air gunners or transferring to the army. At the request of the British government, RAF schools in Canada were among the first to close.

This curtailment of training had little effect on Jarvis, though the instruction of air bombers did decrease dramatically following the November 17 graduation ceremonies of 34 Canadian, British and Polish students from Course 116. By contrast, the school continued to turn out a class of wireless air gunners every two weeks right up until the time the station became inactive in early February 1945. There always was, it seemed, a need for air gunners, who were disappearing at an alarming rate in Bomber Command.

Sights on Jarvis

Lysanders of drogue flight perched on the apron and ready to go. At any given time up to 15 Lysanders were used for target towing at the school.

A well practiced and quickly executed ditching action enabled the pilot and crew of four to escape from their sinking Bolingbroke which crashed into the icy waters of Lake Erie while on a routine training exercise on November 25. The plane, which apparently lost control during a final turn on a gunnery run, flattened onto the water, striking both props. The aircraft bounced, then struck the water again, breaking the fuselage forward of the mid-upper turret. Witnesses to the crash were Selkirk-area farmers Andrew and Marjorie Barnard.

"There was a big splash the first time the plane hit the water," Barnard told a reporter from the *Port Dover Maple Leaf.* "It bounced off the water and when it hit again it broke in two. It was lucky that it did because otherwise the turret gunner might not have been saved. He was trapped but the other boys got him out."[8]

Watching the drama unfold from his lakeshore home, Barnard immediately raced to a small boat and began rowing towards the partly submerged aircraft a mile from shore while his wife telephoned the airport at Jarvis. The crew meanwhile had managed to release a small rubber dinghy from the stowage compartment in the wing, inflate it and clamber aboard. Despite their best efforts to paddle towards shore, they continued to drift east until Barnard, with a good set of oars, managed to reach them and take them in tow.

"They were pretty glad to see me," Barnard confessed. "It was cold on the lake and several of them had been tossed into the freezing water by the crash. One was slightly injured so I placed him beside me, tied their dinghy to my boat and started back."[9]

132

As they neared shore they were joined by Albert Swent and Jack Featherstone in a second, larger boat, along with one of the crash boats from the bombing range. At the Barnard home they were met by the station ambulance and treated by the school's medical officer while a Lysander circled overhead ready to drop an additional life raft if necessary. The crew of five—a pilot, gunnery instructor and three students—were cold, shaken and wet but otherwise in good spirits. Those long, seemingly needless dinghy drills in the lake had apparently paid off.

ROYAL CANADIAN AIR FORCE

JARVIS, Ontario,
27 November, 1944.

Mrs. Andrew Barnard,
Concession #1
Rainham, Ontario.

Dear Mrs. Barnard:

May I express my deep appreciation and also pass on to you the grateful thanks of the crew of the aircraft which was ditched in the lake on Saturday, November 25th, 1944.

Your prompt action enabled us to proceed to the scene with rescue craft and ambulance, which in all probability saved some of the crew members from serious illness and possible drowning.

Again thanking you I remain

Respectfully yours,

(T.H. Mitchell) Flight Lieutenant,
Officer Commanding Flying.
No.1 B&G School,
JARVIS, Ontario.

The letter of thanks sent to Marjorie Barnard from the school. Strangely, her husband Andrew, who rowed singlehandedly to the scene and took a drifting dinghy in tow, is not even mentioned. SHAWN BARNARD

The winter of '44 will long be remembered by residents in that part of Ontario as the worst in recent memory. Beginning December 11 and continuing for more than a week, heavy snow and high winds blocked all roads in the surrounding district, rendering all runways and taxi strips unserviceable. Emergency calls from No. 1 Training Command in Toronto instructed the station to assist the Department of Highways in clearing all main roads and to disregard the runways. No one felt the impact of this decision more, perhaps, than the crew of a lone RCAF Lockheed aircraft which circled the school, apparently in trouble, only to be sent on its way owing to the unplowed state of the airfield.

Sights on Jarvis

On December 15, 200 people—truck drivers, airmen, airwomen, soldiers and civilians— were stranded in the ongoing blizzard that saw more than 100 trucks, cars and US Army vehicles bogged down on Highway 3. At the height of the storm that Friday evening, 11 people were rescued in a horse-drawn sleigh by Robert Dosser, a local farmer. Sgt. George Vinall of Brantford, an airman stationed at Jarvis, braved the howling wind and blinding snow to walk the two-and-a-half miles to get Dosser and the sleigh and guide him back to six stalled cars. Three Toronto women and a Hamilton girl, aided by airmen, walked into the village of Jarvis along the route Vinall took and were in a state of collapse when they arrived. All four had attended a wings parade at the school earlier that day.

In town, they were astounded to find cars stuck in drifts up to their windows. At the main intersection of Highways 3 and 6, 11 transport trucks loaded with wartime cargo and 20 American army vehicles were likewise buried deep in snow, along with a bus bound for Hamilton. At the Jarvis Hotel, they found it packed with servicemen and women, army truck drivers and civilians. All the rooms were booked, forcing the overflow of guests to sleep in the lobby, sprawled on anything available.

Responding to an appeal from the commanding officer at Jarvis, a small convoy of trucks from No. 16 SFTS made its way to town along Highway 6, which by this time had been partially opened by snowplows from the Department of Highways. RCAF airmen and women were loaded up at the hotel and taken back to the Hagersville air school, located just three miles northwest of town. Townspeople took others into their homes. The storm continued all night and by morning the town's lone restaurant was scouring the village for breakfast for its guests.

After 101 hours of continuous plowing by the station's works and buildings section using four snowplows and RCAF snow blower, the situation was momentarily relieved. Then a second blast of winter blocked roads and runways again! Reports of enormous snowdrifts, piled as high as telephone wires in places, poured in from throughout the district. Not until nine days

Civilians and RCAF personnel employed in works and buildings. Front, l-r: LAC William Lovegrove, Cpl. William Templeton, Sgt. Harry Clarke, Sgt. Charles Cocks, F/S Edward Lord, F/L E. Hurry, Sgt. John Wade, Sgt. George MacNeill, Cpl. W.L. Harris, Cpl. Edward Hall, LAC Cecil Hedges. Centre: Charles Winger, Roy Goodwin, John Kraft, Herbert Beck, Archie Bunn, LAC Roger Lortie, AW1 C.F.M. Prangley, LAC F.C. Cole, LAC E.G. Bidnall, Howard Hoover, Emerson Nie, Reg Hedges. Back: Cliff Walker, Bill Hockley, J. Tait, J.C. Swarts, Ivan Nieson, Amos Bartlett, Harry Cox, Milton Henning, John Chapman, Chester Rhora, Claude Nie, Roy Fleming, James Grimster. MICK HENNING

later could a limited flying program be resumed. Much to the despair of locals, it would be early March before many secondary roads in the county could be completely cleared of snow, reopened with the help of RCAF snowplows and bulldozers brought in from the Brennan Paving Company of Hamilton.

With the outcome of the war no longer in doubt that December, G/C McFarlane reflected the sentiments of many in his year-end address to the school. He talked about the growth of the RCAF from a small force of 3,200 personnel in 1939 to its current strength of nearly 200,000 and how Canadians serving in the RCAF had distinguished themselves in all parts of the world. He thanked everyone for the support and cooperation given him the past year and concluded by saying:

Sights on Jarvis

During these years of war we have all had to adjust ourselves to a new way of life. We have developed new interests and made new friends and our former civilian lives have become something only dimly remembered.

The time is coming now when we must become civilians once again, and again pick up the threads which were disrupted by war. For some of you, this will be difficult and will call for the same degree of courage and fortitude which you have shown yourselves to possess. It should be remembered that "Peace hath its victories no less renowned than war" and it is our duty to carry on and to see that this peace, when it comes, will be a lasting one.[10]

Chapter Six
1945 - End in Sight

A senseless tragedy on the evening of January 7 marred the final few weeks of the school's operation. A weather check, needlessly authorized on a night when no bombing could have possibly taken place anyway, resulted in a fatal crash a mile-and-a-half west of the station. On that night P/O Thomas Perley-Martin, in company with two bombing instructors and an aero engine mechanic, took off in Anson 7013 to determine definitely if bombing should be scrubbed.

Almost immediately upon take-off the crew found themselves in trouble, encountering thick fog and icing conditions at 200 feet. Circling back, the 23-year-old pilot radioed for assistance and returned to the field, guided by flares fired by personnel from the catwalk of the control tower. In a desperate attempt to locate the runway, the pilot made three low passes above the field and it was during the last such circuit that the aircraft struck a tree and crashed, killing Perley-Martin and seriously injuring two others. One of the passengers, P/O George Bernhardt, later stated that he and Perley-Martin had become so preoccupied with scraping ice from the windshield with their fingernails, that the pilot had momentarily neglected his altitude when they struck the top of the 25-foot tree.

P/O T.H. Perley-Martin was killed flying in bad weather shortly before the station closed. NAC C144231

Accusations and finger-pointing followed in attempt to discover who made the final decision to fly that night—P/O Perley-Martin or the acting chief instructor of the school, who himself had declined to be part of the crew. Strangely enough, Perley-Martin, as officer commanding night flying, had the authority to overrule the higher-ranked chief instructor but chose not to. Testimony from several witnesses on duty that night indicated the pilot had felt pressured into making the flight, despite the poor weather conditions that had existed all day. Indeed, so questionable was the weather prior to takeoff that lots had been drawn by the bombing instructors to determine who would actually accompany the pilot on the flight! All in all it was an accident waiting to happen and not one of the finer moments in the station's otherwise proud history.

The crash that killed P/O Perley-Martin and injured three others. The pilot became disoriented in fog and icing conditions, striking the top of an apple tree. NORM WILLIAMS

1945 - End in Sight

On January 19 members of Course 95 Wireless Air Gunners completed all ground studies and exams and were posted to No. 4 Bombing and Gunnery School Fingal to finish their flying training. By early February the end was in sight. Several

Chuck Watkinson and Tommy Bull on closing day, February 17, 1945.

bombing and gunnery schools located across Canada had recently closed, with four more, including Jarvis, scheduled to follow later that month. Closer to home, in Haldimand and neighbouring Brant County, service flying training schools at Dunnville and Brantford had already discontinued day-to-day operations.

A final wings parade took place on February 2 when 125 students graduated from Course 96 Wireless Air Gunners and No. 120 Air Bombers. To mark the occasion, wings were presented by the acting chief Polish liaison officer from air force headquarters in Ottawa. As a gesture of friendship and appreciation similar to that expressed to G/C Bell-Irving in September 1943, honourary Polish pilot's wings were awarded to the officer commanding, G/C W.J. McFarlane. Following the ceremony, 83 graduating air bombers were posted to Malton, St. John's or Charlottetown for further navigational training, while the gunners departed on leave prior to reporting to embarkation depot. With the school officially becoming inactive that same afternoon, all training ceased.

Word of the impending closure of the base came as a relief to many. To others, it came all too soon. Within days, many of the more than 1,200 personnel still remaining on the station were transferred to other units, with the balance being discharged or called up for army service under new government regulations. Affected more than anyone, it seemed, were members of the community. The school had forever left its mark in the district where untold numbers of airmen and their families had resided throughout the course of the war, establishing friendships lasting to this day. In its Friday, February 16 edition, the *Port Dover Maple Leaf* summed up community feelings in an editorial:

> Although steeled for the news, it came as a shock to most citizens of this little lakeside town where hundreds, yes thousands of officers and other personnel have made their way home from the day the Jarvis airport first opened. Many officers, non-commissioned officers and other ranks have come and gone throughout the period of operation of the airport, and taking all in all, they've been a grand bunch of fellows whom it has been a pleasure and privilege to have met and who have left in the many homes in Port Dover, a lasting genuine friendship that should never be erased from the happy incidents of the most cruel of all wars. ...Their visit and stay in Port Dover has been the means of making many lasting friendships. Port Dover

Airmen's plot, Knox Presbyterian Church cemetery, September 2, 1945. Temporary crosses, seen here on three of the graves, were replaced soon after by permanent granite headstones courtesy of the Commonwealth War Graves Commission. Flowers on the graves were placed by members of the Jarvis Women's Institute. MYRTLE JOHNSON

is truly proud to have played host to so fine a group of folks as those attached to No. 1 Bombing and Gunnery School. And may you too, delightfully carry happy memories in your hearts of this little lakeside community... [1]

No. 1 Bombing and Gunnery School formally disbanded on February 17, 1945. In the four-and-a-half years the school operated, more than 4,000 bomb aimers and 2,500 air gunners and wireless air gunners from throughout Canada, Newfoundland, the British Empire, Norway, Poland and the United States, received their wings. During that time countless others passed through its gates as mechanics, tradesmen, pilots, instructors, armourers, kitchen staff, civilians, administrative, medical and maintenance personnel. Only a handful remained now, tying up loose ends and providing security while awaiting posting. Looking about the well-kept camp with its paved streets, lush lawns and freshly planted trees, it hardly seemed possible it could be the same place that in 1940 was little more than a sea of mud stirred up by constant rain and the huge contracting equipment brought in for construction purposes.

From those early days comes the story of a Battle pilot and his crew who had just returned to the field after a training flight one winter afternoon in 1941. As they trudged across the tarmac, a visiting newspaper reporter asked the pilot how the flight had gone. Breaking into a smile, the airman explained that just minutes before landing they had encountered a number of Harvards and Yales from the Dunnville flying school, mixing it up and generally indulging in foolhardy aerobatics.

"'They were dog-fighting so I dived in and broke it up,' grinned the Jarvis pilot. 'I thought they might hurt themselves.'"[2]

Epilogue

*I*mmediately upon closure, the aerodrome became the RCAF's No. 404 Reserve Equipment Maintenance Satellite, a facility where aircraft could be temporarily stored until war's end. Despite political wrangling and wire-pulling in high places, hopes that the airport might be retained for peacetime development were quickly dashed when the Crown Assets Corporation began disposing of planes, equipment and other inventory in late 1945. During the next year-and-a-half, dozens of Bolingbrokes stored on the site were scrapped—crushed unceremoniously by bulldozers against the gunnery backstop. At the same time all six hangars and most other buildings were torn down, sold off for lumber and shipped to places as far away as Quebec. For eight years the remaining property, still cluttered with vast stretches of runways, aprons, paved streets and cement hangar floors, was leased to Walpole area farmers then eventually sold to Russell Hare and son Larry.

It was the first of what would be many changes. For 10 years beginning in 1955, the former runways were turned into a racetrack by the British Empire Motor Club of Toronto and given the name Harewood Acres. When the club moved to Mosport in 1965 the track was taken over by members of the London Sports Car Club, who held car races there until 1970. A year later the site was the scene of the International Plowing Match, hosted by Haldimand County and attracting thousands of people between October 12 and 16. In June 1974, Texaco Canada purchased the property and began construction of a

$500 million dollar oil refinery which was sold to Imperial Oil in 1989. Today, only the cement gunnery backstop and tall rows of trees, once planted long the roadways of the school, remain as physical evidence of the former airfield.

However, as a reminder of the sacrifices once made by so many, a commemorative plaque was unveiled on August 21, 1993 near the original main entrance to the airport. Bearing the crest of the Royal Canadian Air Force, it contains a brief history of the base with the names of those who died while serving at the school inscribed on the back. Funding for the plaque was generously provided by Imperial Oil and Wing 412 of the Royal Canadian Air Force Association. Among those attending that day was Haldimand-Norfolk MP Bob Speller; City of Nanticoke mayor Rita Kalmbach; George McMahon Sr., national president of the RCAF Association; the Royal Canadian Legion, Branch 79, Simcoe and members of RCAFA Wing 412, Windsor. Organizers were pleased by the large

Removing buildings took months. Most lumber and siding material was sold to local farmers while the six hangars were dismantled and shipped as far away as Quebec. At least one hangar was reconstructed and served for years as a community hockey arena in southwestern Ontario. GLENN MURPHY

Twenty veterans, many of whom had not returned to Jarvis since the war, attend a plaque unveiling at the former airfield (now Imperial Oil, Nanticoke) on August 21, 1993. KIRK BROWN/HALDIMAND PRESS

number who turned out for the event, and especially the 20 Jarvis veterans who made their way one more time to the place they once called home.

In 1998 Neil Bell, a grade six teacher at Jarvis Public School and lieutenant commander with the Naval Reserve, Canadian Forces, initiated a special day of recognition for the 10 Commonwealth airmen buried in the cemetery directly adjacent to the school yard. In November of that year, all classes participated in a Remembrance Day service which began at the local cenotaph and concluded at the airmen's plot in the Knox Presbyterian Church cemetery. There, as the name of each airman was read aloud, two students stepped forward and laid roses beneath small national flags placed at the base of each headstone. It is the intention of Neil and others like him to make such ceremonies annual events at the school, in honour of those who eagerly joined the air force seeking adventure, but who met death in farmers' fields in southern Ontario instead.

Theirs is a story which must not be forgotten.

Casualties at No. 1 Bombing & Gunnery School

Royal Canadian Air Force

F/O	Bounds,Edwin Jackson (U.S.)	9 December 1941
LAC	Boyd, Joseph Arthur William	19 November 1940
LAC	Burke, Gordon Cragg	18 August 1942
LAC	Burnep, Harold Philip	23 July 1942
AC2	Deebank, Percy John	8 November 1940
Cpl.	Doan, William Cecil	3 January 1943
P/O	Edwards, Gordon Francis*	15 November 1944
LAC	Gray, John Samuel Willis	9 December 1941
P/O	Green, John Howard	31 July 1944
Sgt.	Hawke, Frederick Joseph	9 September 1942
LAC	Hayes, Arthur Raymond	15 June 1943
LAC	Holgate, John Arthur	18 June 1943
LAC	Innes, Frederick Earl	15 June 1943
Sgt.	Jackson, Gordon Bert**	17 September 1941
LAC	Kearny, John Henry	15 June 1943
LAC	Killick, Ronald William Gillingham	9 September 1942
LAC	Kirkby, Warren Milton	18 August 1942
LAC	Lefurgey, John Alfred	18 August 1942
Sgt.	McCrank, Robert Neil	9 September 1942
AC2	Morehouse, Ira Bertrum**	30 October 1941
F/O	Norbury, Edward Blake	15 June 1943
P/O	Perley-Martin, Thomas Henry	7 January 1945
F/O	Poole, Maxwell Boyer	18 August 1942
LAC	Reed, Archibald Campbell	18 August 1942
LAC	Samuel, Hector Patrick	15 June 1943
LAC	Schwartz, Solwyn Saunders (U.S.)	23 July 1942
LAC	Smith, James Alexander	15 June 1943
F/S	Troutbeck, George Roland (NZ)	4 January 1943
LAC	Winfield, George Norman	9 November 1942

*Died while on leave   **Motor vehicle accident

Royal Air Force

LAC	Barber, Frederick George	9 December 1941
LAC	Best, Gordon Cooper	3 July 1944
Sgt.	Wade, Norman	9 November 1942
LAC	Waller, Robert	3 July 1944
LAC	Williams, John Strainer	23 July 1942

Royal Australian Air Force

LAC	McNabb, Raymond	6 July 1941
P/O	Slater, Kenneth	23 July 1942
LAC	Taggart, Charles	6 July 1941
F/S	Watts, John Bradford	3 July 1944

Royal New Zealand Air Force

Sgt.	Whitehead, John William	18 August 1942

Civilian

	Bullock, Harry**	7 May 1941

Burials at Knox Presbyterian Church, Jarvis

Troutbeck, George Roland (RCAF - New Zealander)
Best, Gordon Cooper (RAF)
Wade, Norman (RAF)
Waller, Robert (RAF)
Williams, John Stainer (RAF)
McNabb, Raymond (RAAF)
Slater, Kenneth (RAAF)
Taggart, Charles (RAAF)
Watts, John Bradford (RAAF)
Whitehead, John William (RNZAF)

All others were returned home for burial except LAC Frederick George Barber, RAF. At his girlfriend's request, he was re-interred at Woodlawn Cemetery, Guelph.

Appendix 2
Chronology
1940

April 4 – The *Jarvis Record* announces Jarvis as the location for one of two Bombing & Gunnery Schools to be developed immediately under the British Commonwealth Air Training Plan.

April 11 – Ontario contractors begin to arrive on the site including Johnson Brothers and Schultz Construction, Brantford; Sterling Construction, Windsor; Frid Construction, Dufferin Paving and Crushed Stone, Grant Construction and Colas Roads, Toronto.

May 1 – Leases signed between property owners and Federal Government for 600 acres required for airport.

June 20 – Acting Minister of National Defence and Minister of Defence for Air allots $770,000 for Jarvis Bombing & Gunnery School.

July 25 – Advance party of airmen arrive under the direction of S/L R.A. Cameron.

August 9 – First six aircraft taken on strength.

August 18 – Thirty-nine pupil air observers reach the school, followed a week later by the first course of air gunners.

August 24 – G/C G. E. Wait of Ottawa assumes command of the station.

September 12 – First practice bomb hits 25 yards from No. 1 target at Peacock Point.

September 29 – First class of air observers graduate from the school.

October 28 – First air gunners to graduate in Canada under BCATP receive wings from Air Vice Marshal L.S. Breadner, DSC, Chief of the Air Staff.

November 8 – AC2 P.J. Deebank becomes school's first casualty when he drowns retrieving a drogue which fell into the lake at Hoover's Point.

November 19 – LAC J.A.W. Boyd is killed instantly when he accidently steps into revolving propeller.

1941

February 17 – Eleven Norwegians arrive for training followed a month later by students from Great Britain, Australia and New Zealand.

March 25 – Head producer of Warner Brothers studio visits Jarvis in preparation for the motion picture *Captains Of The Clouds*, released in 1942.

April 22 – A Harvard trainer modified for bombing and gunnery practice is flown in to the station to be tested for possible use at RCAF bombing and gunnery schools.

May 7 – H. Bullock, civilian, is killed at night by RCAF station wagon.

June 7 – Wings presented to graduates of Jarvis and Fingal by Air Marshal W.A. "Billy" Bishop, at No. 1 RCAF Manning Depot, Toronto.

July 6 – Australian students C. Taggart and R. McNabb become school's first fatalities as a result of a plane crash.

August 27 – HRH the Duke of Kent, visits station.

September 17 – Sgt G.B. Jackson, on staff of the ground instruction school, is killed in an automobile accident while returning from leave.

October 21 – In preparation for night bombing, staff pilots begin regular night flying on the station.

October 30 – AC2 I.B. Morehouse is killed and another airman seriously injured in a motorcycle accident near Port Dover.

November 30 – Night bombing practice, the first ever planned in Canada, is introduced at Jarvis.

December 8 – The United States declares war on Japan. Fewer than one third of the 100 Americans serving at Jarvis resign from the RCAF in order to enlist in their own armed forces.

December 9 – Three flyers die when two aircraft collide west of Fisherville.

December 20 – First class of wireless/operators to graduate from No. 4 Wireless School, Guelph, receive their AG badge at Jarvis.

1942

February 15 – Wings presentation carried live over CBS radio network.

February 20 – The 1000th air gunner graduates from the school.

March 20 – G/C G.E. Wait is succeeded by W/C W. Hannah as Commanding Officer.

April 27 – First members of the RCAF Women's Division arrive at Jarvis.

June 21 – 167 air cadet officers begin a two week course in various phases of air force routine.

July 23 – Bolingbroke 9113 plunges into Lake Erie at Peacock Point, claiming four lives.

July 31 – G/C A.D. Bell-Irving, MC, assumes command of the station from W/C W. Hannah.

August 18 – Six flyers are killed in mid-air collision.

August 29 – First class of air bombers (formerly air observers) graduate from Jarvis.

September 6 – Student air observer scores direct hit on motorized practice boat during moving target practice.

September 9 – Bolingbroke crashes at Clanbrassil killing three.

October 25 – School sets a new RCAF gunnery record with 169 exercises completed of 200 rounds each.

October 27 – Party of volunteers help extinguish flames on freight train burning out of control at Nelles Corners. F/O J.F. Williams and Sgt R.A. Picard receive the George Medal and British Empire Medal respectively.

November 9 – A pilot and mechanic die in plane crash on the Six Nations Reservation.

1943

January 3 – Bolingbroke 9964 loses power on take-off and crashes, killing two of the four men onboard.

February 9 – Jarvis sends six Bolingbrokes under the command of F/L M.T. McKelvey to instruct high-level bomb aimers at Paulson, Manitoba.

May 22 – Station medical officer and marine section assist the crew of a Hagersville Anson forced down in the lake near Long Point.

June 5 – Jarvis establishes new RCAF 24-hour training record with 127 hours flown by Bombing Flight, 197 exercises completed and a total of 1241 bombs being dropped.

June 12 – Air Marshal W.A. Bishop presents wings to members of Course 54, wireless air gunners.

June 15 – Seven flyers lose their lives when two Ansons on a routine bombing exercise collide at 2,000 feet above Oneida Township.

September 1 - Station receives a visit from W/C G.P. Gibson, during a cross Canada tour. Gibson led the "Dam Busters" raid in May, 1943.

September 18 – G/C A.D. Bell-Irving, MC, presented with Honourary Polish pilot's badge by G/C S. Sznuk of the Polish Air Force.

November 1 – W/C W. Peace, DFC, assumes temporary command of the station; G/C Bell-Irving posted to the central flying school at Trenton.

December 31 – Station reaches full capacity with 1,857 men and women attached to the base, including 147 civilians.

1944

February 5 – G/C W.J. McFarlane assumes command.

May 3 – Station awarded minister's efficiency pennant.

June – New RCAF synthetic trainer is installed on the Air Ministry Bombing Teacher at the school.

June 7 – Simcoe and area public school students donate an airplane to the RCAF. No. 1 B &G School accepts on behalf of the air force.

July 3 – An Anson with three men onboard catches fire while returning from a bombing flight and crashes in flames near Port Dover.

July 31 – Lysander 2317 plunges to the ground at Rainham Centre killing J.H. Green. The drogue operator, S.W. Garland, parachutes to safety.

August 16 – Airport opens its doors to first-ever Jamboree and Sports Day. More than 6,000 civilians and service personnel attend.

November 15 – P/O G.F. Edwards, an officer on staff at the school, dies in hunting accident while on leave.

November 25 – Local man rows to scene of crash in lake and tows drifting dinghy with five airmen to shore.

December 11 – Blizzard of '44 begins, blocking all roads and shutting down flying operations at the school for days.

1945

January 7 – P/O T.H. Perley-Martin of Vancouver is killed when Anson on weather check clips the top of a tree in poor visibility.

January 11 – Word is received from Air Force Headquarters that No. 1 Bombing & Gunnery School will close on February 17, 1945.

February 2 – Final wings parade takes place with the graduation of 42 wireless air gunners and 83 air bombers. School officially becomes inactive later that same day and all pupil training ceases.

February 5 – Several buildings closed. The drill Hall is taken over for storage of barrack equipment and telephones disconnected.

February 17 – Station is formally disbanded.

Endnotes

Chapter One: 1940 - War Comes to Walpole

1. "Memorial Service Held Honoring Ontario Pilot", unidentified newspaper clipping, April 28?, 1940.
2. "Old Timers" by AC1 T.L. Mather in *The Fly Paper*, February 1943.
3. "Airmen Seek Accommodations", *The Jarvis Record*, August 8, 1940.
4. Daily Diary, No. 1 Bombing and Gunnery School, August 5, 1940 through September 16, 1940.
5. "First Class of Gunners Graduate at Air School", *Simcoe Reformer*, October 28, 1940.
6. "Jarvis Aftermath ... or 'The Dismal Dozen'" by Eric Cameron, *Short Bursts*, September 1993, p. 13.

Chapter Two: 1941 - Setting the Standard

1. "Reflections of a Bombing Pilot - F/L Burton Recalls His First Year At Jarvis," *The Fly Paper*, date unknown.
2. Interview with Harold Waterbury, Fisherville, Ontario, May 1975.

Chapter Three: 1942 - The Terrible Summer

1. "Low Flying Stunting Planes Endanger Life Says Airport CO", *Port Dover Maple Leaf*, September 4, 1942.
2. "Four Airmen Die As Plane Crashes At Bombing Range," *Simcoe Reformer*, July 27, 1942.
3. Solwyn Schwartz to Jeanne Schwartz, letter dated 22 July 1942.
4. Flying logbook of Norman Davidson, August 5, 1942.
5. RCAF Investigation H.Q. File No. 1100-8207, August 19-21, 1942.
6. "In Remembrance", *The Fly Paper*, August 20, 1942.
7. Interview with Margaret Roberts, Clanbrassil, July 1997.
8. "Remustered Ground Crew Received Wings At Jarvis Saturday," *Port Dover Maple Leaf*, October 2, 1942.
9. RCAF Investigation H.Q. File No. 1700-2325, p.6.

Chapter Four: 1943 - In Full Swing

1. *Mentioned in Dispatches*, September 1943. (Department of National Defence publication).
2. Interview with Tom Lawrence, Fort Erie, August 4, 2003.
3. "Billy Bishop Presents Wings At Jarvis School," *Simcoe Reformer*, June 14, 1943.
4. RCAF Investigation H.Q. File No. 1700-7339.
5. "Aircrew A Great Team, Says V.C.", *The Fly Paper*, September 1943.
6. "Jarvis C.O. Presented With Polish Air Badge," *Simcoe Reformer*, September 20, 1943.

7. Interviews and correspondence with Enid Blume, April-May 2002.

8. *Ibid.*

9. *Ibid.*

Chapter Five: 1944 - Leveling Off

1. *Port Dover Maple Leaf*, January 14, 1944.

2. "Officer Receives DFM," *The Fly Paper*, April 1944.

3. Newspaper clipping, n.d., n.p.

4. Interview and correspondence with Irene (Osborne) Miller, Dunnville, Ontario, August 2002.

5. RCAF Investigation H.Q. File No. 1700-2317, p. 2.

6. "Expand Pilot's Training to Two Types of Aircraft," *Globe & Mail*, July 26, 1944.

7. "Airmen Qualify At Hagersville In Double Course," *Hamilton Spectator*, August 26, 1944.

8. "Five Airmen Rescued From Lake Erie Crash," *Port Dover Maple Leaf*, December 1, 1944.

9. Newspaper clipping, n.d., n.p.

10. *The Fly Paper*, December 1944.

Chapter Six: 1945 - End in Sight

1. "Jarvis Bombing & Gunnery School Closes Saturday", *Port Dover Maple Leaf*, February 16, 1945.

2. "Not One Moment Is Being Lost At Great Field Near Jarvis", *Hamilton Spectator*, January 30, 1941.

References

Books

Allison, Les and Harry Hayward. *They Shall Grow Not Old: A Book of Remembrance.* Commonwealth Air Training Plan Museum, Brandon, 1991.

Bishop, William Arthur. *The Courage of the Early Morning.* David McKay Company Inc., New York, 1965.

The Canadians at War, 1939/45, Vol. 1. Reader's Digest Association (Canada) Ltd., 1969.

Colombo, John Robert, *Colombo's Canadian References.* Toronto: Oxford University Press, 1976.

Conrad, Peter C. *Training For Victory: The British Commonwealth Air Training Plan in the West.* Western Producer Prairie Books, Saskatoon, 1989.

Dosser, Shirley. *Nanticoke Through the Years.* Norfolk Historical Society, Simcoe, Ontario, 1990.

Douglas, W.A.B. *The Official History of the Royal Canadian Air Force. Vol. II: The Creation of a National Air Force.* University of Toronto Press, 1986.

Dunmore, Spencer. *Wings For Victory.* McClelland & Stewart Inc., Toronto, 1994.

Hatch, F.J. *Aerodrome of Democracy: Canada and the British Commonwealth Air Training Plan 1939-1945.* Directorate of History, Ottawa, 1983.

Hewer, Howard. *In For A Penny, In For A Pound.* Stoddart Publishing Co. Limited, Toronto, 2000.

Lennon, Mary Jane and Syd Charendoff. *On the Homefront.* The Boston Mills Press, Erin, Ontario, 1981.

MacDonald, Cheryl. *Port Dover: A Summer Garden.* Port Dover Board of Trade, Port Dover, Ontario, 2001.

Milberry, Larry. *Sixty Years.* Canav Books, Toronto, 1984.

Milberry, Larry and Hugh A. Halliday. *The Royal Canadian Air Force at War 1939-1945.* Canav Books, Toronto, 1990.

Olmstead, Bill. *Blue Skies.* Stoddart Publishing Co. Limited, Toronto, 1987.

Ozorak, Paul. *Abandoned Military Installations of Canada Volume I: Ontario.* Paul Ozorak, Ottawa, 1991.

Post, Bill. *Jarvis and the American Airlines Experience.* Jarvis Library Guild Book Committee, Jarvis, Ontario, 2003.

Williams, James N. *The Plan: Memories of the BCATP.* Canada's Wings, Inc., Stittsville, Ontario, 1984.

Journals, Magazines and Newspapers

Brantford Expositor

Haldimand Press

Hamilton Spectator

Jarvis Record

London Free Press

The Fly Paper (No. 1 B&GS)

The Sky Writer (No. 6 SFTS)

Port Dover Maple Leaf

Short Bursts

Simcoe Reformer

Schweyer, Robert. "Jarvis: The First Three Years" in *CAHS Journal*. Canadian Aviation Historical Society, Willowdale, Ontario, 1978.

Primary Sources

Personal correspondence and interviews with former students and residents of the area surrounding No. 1 B&GS

Unit Diary, No. 1 B&GS, RCAF, Jarvis.

Unit Diary, No. 6 SFTS, RCAF, Dunnville.

Unit Diary, No. 16 SFTS, RCAF, Hagersville.

Photo Credits

Most of the photographs appearing in this book are from personal collections and are credited to the owners, when known. Photos not accompanied by a credit are from the collection of the author. Additional sources (and the abbreviations used in the credits) include those listed below:

Commanding Officer, Canadian Forces Photo Unit, Department of National Defence (DND), Ottawa.

National Archives of Canada (NAC), Ottawa.

London Free Press Collection of Photographic Negatives, University of Western Ontario Archives, London.

Canadian Harvard Aircraft Association (CHAA), Tillsonburg.

Jarvis Public Library, Jarvis.

Canadian Warplane Heritage Museum (CWH), Hamilton.

No. 6 SFTS Reunion Association, (No. 6 SFTS), Dunnville.

Discharged Personnel Records, Department of Defence, Royal Australian Air Force, (RAAF) Canberra, Australia.

Personal Names Index

Knipe, Harry *40*
Kraft, John *135*

- *L* -

Lawrence, Tom 96
Lefurgey, John 75, 77, 145
Lord, Edward *135*
Lortie, Roger *135*
Lovegrove, William *135*

- *M* -

Mack, J. W. 95
MacKelvie, James 76
MacNeill, George *135*
Makey, Elvin 50
Marshall, Brenda 36
Martin, Jack 43
Mayhew, L.G. 82
McCrank, Robert 77, 78, 79, 145
McFarlane, W.J. 116, 121, 122, *129*, 135, 139, 150
McKelvey, M.T. 90, 149
McLean, Lloyd 87,88
McMahon, George Sr. 143
McNabb, Ray 41, 146, 148
McRae, John 117, 118
Miller, Emerson 17
Miller, James 17,18
Milner John *80*
Montigny, Lawrence 87
Montrose, J. Wally 126
Moon, Ian *10*
Moore, Harold 41
Moorehouse, H. C. 51
Morehouse, Ira Bertrum 145, 148
Morgan, Dennis 36
Morris, Stan 54
Moyes, R.K. 80
Murray, Bruce *86*
Mussell, Bill 92, 93

- *N* -

Nash, Gerald 129
Nie, Claude *135*
Nie, Emerson *135*

Nieson, Ivan *135*
Norbury, Edward *102*, 103, 145
Norfolk, Ken *63*

- *O* -

O'Hara, Roderick 78
Ohland, Harvey 83
Olmstead, Bill 64

- *P* -

Paton, Ila *123*
Paton, Jack *40*, *123*
Patriarche, V. H. 65
Peace, William 111, 149
Pearce, F.C. 21,22
Pearcy, C.G. 36
Perley-Martin, Thomas 137, *138*, 145, 150
Picard, Ray 81,82, 149
Poole, Maxwell 75, 145
Potter, Ted *91*, 105
Prangley, C.F.M. *135*

- *R* -

Raine, Norman Reilly 36
Reed, Archibald 75, 76, 145
Reistad, Ole 131
Rhora, Chester *135*
Richthofen, Manfred Von 74
Roberts, Margaret 78
Russell, Keith 51
Ryerse, Edmon *90*, 97
Ryerse, Robert *124*

- *S* -

Samuel, Hector *102*, 103, 145
Schneider, Enid *51*
Schwartz, Frank 6
Schwartz, Jeanne 67, 68
Schwartz, Solwyn *66*, 67, 68, 145
Scott, G.O. 80
Sibley, George 87, 88
Slater, Kenneth *66*, 67, 146
Smith, James *102*, 103, 145
Speller, Bob 143
Storey, D.S. 79

Struthers, Jack 54
Swarts, J.C. *135*
Swent, Albert 133
Sznuk, S. 110, 149

- *T* -

Taerum, Terry 108
Taggart, Charles 41, 146, 148
Tait, J. *135*
Templeton, William *135*
Thompson, John *39*
Townley, W.B. 116, 117
Troutbeck, George 87, *88*, *89*, 145
Tunstall, J.A. 82

- *V* -

Vinall, George 134

- *W* -

Waddell, Duke 37
Wade, John *135*
Wade, Norman 83, 146
Wait, George E. 25, 26, *43*, 55, 61, 147, 148
Walker, Cliff *135*
Waller, Robert 126, 146
Waterbury, Harold 50, 51
Watkinson, Chuck *139*
Watts, John Bradford *126*, 146
Whitehead, John Wm. *75*, 76, 146
Whitside, Bill 121
Whitside, Bruce 121
Wilkins, F.S. 115
Williams, Enid 112, *113*, 114
Williams, Jack 81,82, 149
Williams, John 67, 146
Winfield, George 77, *78*, 83, 145
Winfield, Lenore 77
Winger, Charles *135*

- *Z* -

Zimmerman, Herbert 126

General Index

About the Author

Rob Schweyer was born and raised on a farm near Fisherville, Ontario and received a diploma in radio and television broadcasting from Niagara College, Welland, in 1973. Although he worked as a commercial producer at CHML radio in Hamilton for 14 years, his lifelong fascination with airplanes led him in other directions. Rob was employed for eight years as assistant curator at the Canadian Warplane Heritage Museum, Mount Hope and served as public relations officer for the Hamilton International Air Show from 1980 to 1982.

Author of numerous magazine and newspaper articles about Haldimand County's wartime air schools, Rob has also acquired a large collection of artifacts relating to them. He resides in Jarvis with wife Cindy, daughter Sarah and son Matthew.